S
494.5
B563
A36
1993

NABC REPORT 5

Agricultural Biotechnology: A Public Conversation About Risk

EDITED BY JUNE FESSENDEN MACDONALD

WITHDRAWN
FROM LIBRARY
MAR 15 1999
NOVARTIS
CROP PROTECTION

published by

NATIONAL AGRICULTURAL BIOTECHNOLOGY COUNCIL
ITHACA, NEW YORK 14853-2703

The National Agricultural Biotechnology Council provides an open forum for the discussion of issues related to the impact of biotechnology on agriculture. The views presented and positions taken by individual participants in this report are their own and do not necessarily reflect the views or policies of the NABC.

NABC grants permission to copy the overview, workshop summaries, concluding remarks and workshop texts. Permission to copy individual presentations is retained by the authors.

Copying of this report or its parts for resale is expressly prohibited.

Additional copies are available for $5
please make checks or purchase orders payable to:

NABC / BTI
159 Biotechnology Building
Cornell University
Ithaca, NY 14853–2703

© 1993 NABC All rights reserved.
ISBN: 0–9630907–3–9
Library of Congress Catalog Card Number 93–086317

Photos by John Underwood, Purdue University Photographic Services except photo of Jerry Bishop on page 107 which was taken from a videotape of NABC 5 by Purdue University Agricultural Communication Services.

Printed by Braun and Brumfield, Ann Arbor, Michigan
♲ printed on recycled paper

Reports by NABC:

NABC REPORT 1, Biotechnology and Sustainable Agriculture: Policy Alternatives (1989)

NABC REPORT 2, Agricultural Biotechnology, Food Safety and Nutritional Quality for the Consumer (1990)

NABC REPORT 3, Agricultural Biotechnology at the Crossroads: Biological, Social and Institutional Concerns (1991)

NABC REPORT 4, Animal Biotechnology: Opportunities & Challenges (1992)

Occasional Papers by NABC:
Ethics and Patenting of Transgenic Organisms (1992)

Labeling of Biotechnology Products (1993)

National Agricultural Biotechnology Council

The National Agricultural Biotechnology Council is a consortium of not–for–profit agricultural research and educational institutions established in 1988.

Member Institutions

- Boyce Thompson Institute
- Cornell University
- Int'l. Service for the Acquisition of Agri-Biotech Applications
- Iowa State University
- Michigan State University
- North Carolina State University
- The Ohio State University
- Oregon State University
- Purdue University
- Rutgers University
- The Texas A&M University System
- Tufts University
- University of California, Davis
- University of Georgia
- University of Guelph
- University of Missouri, Columbia
- University of Nebraska, Lincoln
- University of Saskatchewan

NABC's principal objectives are to:

- Provide an open forum for persons with different interests and concerns to come together to speak, to listen, to learn and to participate in meaningful dialogue and evaluation of the potential impacts of agricultural biotechnology.
- Define issues and public policy options related to biotechnology in the food, agricultural and environmental areas.
- Promote increased understanding of the scientific, economic, legislative and social issues associated with agricultural biotechnology by compiling and disseminating information to interested people.
- Facilitate active communication among researchers, administrators, policymakers, practitioners and other concerned people to insure that all viewpoints contribute to the safe and efficacious development of biotechnology for the benefit of society.
- Sponsor meetings and workshops and publish and distribute reports that provide a foundation for addressing issues.

ACKNOWLEDGMENTS

The success of the fifth annual National Agricultural Biotechnology Council (NABC) meeting resulted from the hard work and careful planning of the Organizing Committee at Purdue University, the host institution: cochairs Peter E. Dunn, Director, Purdue University Biotechnology Institute, and Marshall A. Martin, Director, Center for Agricultural Policy and Technology Assessment, with J. Alfred Chiscon Jr., Biological Sciences; David A. King, Agricultural Communication Services; Roger P. Maickel, Pharmacology and Toxicology; Lilly–Marlene Russow, Philosophy; Ronald L. Somerville, Biochemistry; and Robert D. Waltz, Indiana Department of Natural Resources. NABC gratefully acknowledges the help of Jean Rosscup Riepe, Agricultural Economics, Purdue University, during the planning of the meeting and the production of this report, and her editing of the videotapes of NABC 5. Also recognized is the important role played by the workshop facilitators: Gary Wagenheim, leader; Howard Borck, Bill Krug and Dan Lybrook, Organizational Leadership and Supervision; and John Trott, Office of Agricultural Resources and the workshop rapporteurs: Laura Hoelscher, Agricultural Communication Services; Brent Ladd,Animal Science; Jean Rosscup Riepe, Agricultural Economics; and Patrick Stewart, American Farmland Trust.

Special recognition goes to the efficient conference service provided by Kathy F. Hyman and her team as well as the many services of Agricultural Communication Services, especially to video cameramen Rick Dodson, Michael M. Kerper and Chris Sigurdson, to photographer John Underwood for his photos of NABC 5 participants, many of which we used in this report and to Dave King for "the communication wheel."

Finally, very special thanks go to Kate O'Hara, NABC Graphics/Production Coordinator for the overall design of this report and her diligent oversight of its production.

June Fessenden MacDonald
Executive Director, NABC
Editor, NABC Report 5

PREFACE

The 1993 National Agricultural Biotechnology Council meeting witnessed the "growing up" of NABC as the leadership passed from Ralph W. F. Hardy, President, Boyce Thompson Institute and cofounder of NABC in 1988, to Bill R. Baumgardt, Director, Agricultural Research Programs, Purdue University and NABC Chair for 1993–94.

Ralph Hardy, NABC Chair from 1988 to 1993, has left his mark on this organization as it has grown under his enlightened and thoughtful leadership from four agricultural research and educational institutions (Boyce Thompson Institute, Cornell University, Iowa State University and University of California, Davis) to eighteen U.S. and Canadian institutions. His vision of a neutral "playing field" where anyone and everyone could come to speak, to listen and to learn is now realized annually. NABC 5, *A Public Conversation About Risk*, is representative of how much can be learned, better understood, productively discussed when people with different views come together in a neutral forum to openly and freely speak about their specific concerns and to work with one another to establish common ground, understand areas of disagreement and, where possible, bring forth mutually agreed upon recommendations.

Since the first laboratory experiment in agricultural biotechnology, there have been questions, discussions and disagreements about "the risks." The fifth annual NABC meeting focused on bringing together and facilitating conversations on risk and agricultural biotechnology among the many stakeholders. Participants entered into dialogues, formal and informal, about technical assessments and personal perspectives of risk, on public perceptions and values of both risks and benefits, and about issues of communication and who to trust. It became obvious to participants that the technical is so intertwined with social issues and values that responses to new agricultural biotechnologies should not be discussed and cannot be understood separate from one another.

This volume provides the reader with many perspectives about how risk is and/or should be assessed. The lively discussions in the workshops and the conversations in the hallways among NABC 5 participants with diverse viewpoints resulted in surprising consensus in the conclusions reached and recommendations presented at the end of the meeting. Hopefully, this report will provide an incentive for the reader to enter into conversations with those who have different perspectives on risk and agricultural biotechnology. Open dialogue will only improve understanding of various viewpoints and should provide a foundation for addressing concerns about agricultural biotechnology.

CONTENTS

PART I

Agricultural Biotechnology: A Public Conversation About Risk

Overview

Peter E. Dunn
Entomology;
Director, Purdue University
Biotechnology Institute
(pictured on left)
and
Marshall A. Martin
Agricultural Economics;
Director, Center for
Agricultural Policy and
Technology Assessment
(pictured on right)

Life involves choices. Many of the choices we make require an assessment of the potential benefits and risks associated with our choices. This is particularly true regarding biotechnology. The major objective of the fifth annual meeting of the National Agricultural Biotechnology Council (NABC) was to exchange understanding of the risks associated with agricultural biotechnology and how these risks are perceived and assessed by various segments of society. NABC 5, *A Public Conversation About Risk,* achieved this objective using the NABC format of an open forum for participants to speak, to listen and to learn. Differing views on risks of agricultural biotechnology were expressed and discussed by over 130 agricultural biotechnology stakeholders including consumers; farmers; environmentalists; industry scientists and administrators; journalists; ethicists; government agency professionals; and academic researchers; educators; and administrators.

The meeting topic and the dates for NABC 5 were rather timely, coinciding with the release of Steven Spielberg's blockbuster science–fiction movie *Jurassic Park* which raised questions about the potential dangers of imagined genetic engineering gone awry. Also, the Food and Drug Administration (FDA) issued, in May 1992, a request for comments on labeling of food products that are biotechnologically based. In addition, the risks of another agricultural technology, pesticides, were being questioned in a National Academy of Sciences report on pesticides in the diets of infants and children since data are not available.

The NABC open forum on the risks of agricultural biotechnology was structured around a roundtable discussion, seven invited plenary speakers, and four concurrent workshops. Each workshop had two cochairpersons who also served as roundtable discussants. The backgrounds of the plenary speakers and workshop cochairs were as diverse as those of the participants.

ROUNDTABLE: A PUBLIC CONVERSATION ABOUT RISK

Marshall A. Martin, NABC 5 organizing committee cochair, moderated a thought–provoking, and often lively and spirited, opening to the NABC 5 meeting. The ideas conveyed in the roundtable set the stage for the remainder of the meeting. (Part IV, page 115, contains a complete transcript of the roundtable discussion.) Different scientific and personal views were expressed about the risks and benefits associated with various technologies, such as alternative modes of travel, consuming dairy products from cows treated with bovine somatotropin (bST) and eating genetically engineered tomatoes. These specific products of biotechnology were selected to draw out a number of issues—such as the role facts, values and emotions play when assessing risk—that were addressed later by the invited speakers and in the workshops. The discussion about the Flavr Savr™ tomato, *Bt* tomato and bST milk, illustrated how individuals can have different tastes and preferences for food, can differ frequently in their environmental concerns, can desire varying levels of information about food products, can have food allergies that might require labeling of transgenic foods, and can display substantial diversity in their knowledge and understanding of agricultural biotechnology. Hence, both technical and other risks and benefits must be considered, though neither is easy to quantify.

PLENARY SESSIONS

During the first full day of the meeting, participants heard presentations from five invited speakers.

Risk Assessment and Communication

Roger A. Balk, an ethicist at the Royal Victoria Hospital in Montreal, spoke on the topic of "Public Values and Risk Assessment." Dr. Balk caught the audience's attention by recalling the story of Henny Penny who concluded, from being hit on the head by an acorn, that the sky was falling—jumping to conclusions that there is a risk without the relevant facts can lead to different, but real risks. Drawing examples from the field of medicine, Dr. Balk described a three–tiered system which would allow advanced technology to be supported while protecting those who would be its beneficiaries. Key elements of this system included: 1. the principle of informed consent; 2. the requirement that all applications for support of new technology be subjected to scientific and statistical cost/benefit analysis; and 3. the development of a process through which principles may be combined with data to guide regulation of the products of new technology. Dr. Balk applied this system to some current, controversial agricultural biotechnologies—bST and Flavr Savr™ tomato.

The next speaker, Sharon Dunwoody, addressed the subject of "Telling Public Stories About Risk." Drawing from her own experiences in dealing with the discovery that her basement may contain asbestos, Dr. Dunwoody, a professor of journalism and mass communications, discussed how people choose information channels when faced with a risky situation. These channels are used to learn about risk as well as to decide how to respond to that risk. She concluded that, given an array of information channels, individuals choose different channels to help them make decisions about different dimensions of a risk. The cost of a channel is also a choice factor. While mass media may be an inexpensive, easily accessible, and important source of information about a risk, persons seeking guidance on appropriate responses to a recognized risk prefer the channel of personal counselors who can take their individual situation into account.

Risk Assessment and Public Perception

During the second plenary session, Roy L. Fuchs spoke on "Risk Assessment: A Technical Perspective." Dr. Fuchs, a Monsanto scientist, presented the perspective of the regulated agricultural biotechnology industry. He began with an overview of the regulatory authority of the various federal agencies and then summarized the approach taken by Monsanto to ensure the safety of one of its biotechnology products: genetically modified potatoes expressing resistance to the Colorado potato beetle. According to Dr. Fuchs, Monsanto's guiding principles were: 1. to establish that the modified potatoes were "substantially equivalent" to the unmodified and widely consumed Russet Burbank potato; and 2. to confirm the environmental, human, and animal safety of the protein products encoded by the two genes introduced via genetic engineering. During the course of his talk, Dr. Fuchs provided numerous examples of the kinds of technical data gathered to evaluate the safety of a genetically modified food.

In his presentation titled "Public Perceptions of the Benefits and Risks of Biotechnology," Dr. Thomas Hoban, a sociologist at North Carolina State University, described results from an extensive national telephone survey and subsequent focus groups. In reviewing conclusions from the survey, Dr. Hoban emphasized that the public is generally optimistic about the potential benefit from and safety of biotechnology in agriculture and desires more information about agricultural biotechnology. However, when specific applications were discussed, considerable variability in perception was uncovered regarding: 1. the public's confidence in sources of information and regulations; and 2. a number of concerns over certain types of products. Survey results indicate that people are more likely to accept plant than animal applications of biotechnology. Views about the morality of biotechnology are very important. Also, respondants want to play a greater role in decisions about biotechnology—he suggested that surveys provide a cost–effective and systematic mechanism for public participation.

Following the Thursday evening dinner, Jerry Bishop, deputy news editor for science with the *Wall Street Journal*, addressed the topic of "Communicating with the Public about Risk." Mr. Bishop recalled the continual stream of risk–associated issues which appears in the news each day. He noted that scientists often criticize journalists for irresponsibility in giving credence to issues, spokespersons, and data which have not been reviewed and evaluated through the critical eye of the scholarly peer–review process. He pointed out that it is often because an issue is controversial that it is of interest to the public and newsworthy. Mr. Bishop reminded the audience that newspapers and other print media must select articles with an eye attuned sharply to what the public wants to read, not what the public should read, if they want to continue publishing.

Risk Assessment and Public Perspective

In the opening plenary session the next morning, Dr. David MacKenzie spoke on "Regulatory Risk Assessment: A View from the Potomac." A U.S. Department of Agriculture (USDA) administrator, Dr. MacKenzie provided an overview of current biotechnology regulation and reviewed the underlying process of risk analysis. In his discussion of risk analysis, Dr. MacKenzie defined such terms as hazard, risk, risk analysis, risk assessment, risk management, risk characterization and risk communication. He emphasized that the science (risk assessment) and the policy/decisionmaking (risk management) must be kept separate. Dr. MacKenzie also discussed issues regarding the Federal Coordinated Framework for Regulation of Biotechnology. He explained how it has been difficult to fit into the Coordinated Framework the regulation of agricultural research and product development because this was a previously unregulated area (as opposed to pharmaceuticals). He indicated that, while regulatory gaps exist, the agencies have done a good job of incorporating agricultural regulation and have worked well together. Although the Coordinated Framework focuses biotechnology regulation on the product rather than the process, the issue of how to identify a hazard before the fact has not been resolved. Thus the process by which a specific product is made may need to be considered as the safety of the product is reviewed. Dr. MacKenzie discussed the future challenges for biotechnology regulation. Educating the public about the scientific facts of biotechnology and reaching consensus by reconciling conflicting worldviews about biotechnology through ongoing dialogue are especially important.

The final plenary speaker was Mr. Will Erwin, an Indiana farmer, who spoke on "Risk Assessment: A Farmer's Perspective." In a warm and thoughtful presentation, Mr. Erwin described the farmers' perspective on new technology in general, characterizing farmers as informed risk–takers. He went on to discuss how farmers' concerns are often dichotomous. On the one hand, farmers exhibit personal concerns over the safety of agricultural practices and products for the environment and the public. On the other hand, farmers also have business concerns over regulatory requirements, profitability, the effect of technology on the structure of the industry, and public demand for products. Mr. Erwin emphasized the complexity of the technological issues involved and the need for more information presented clearly to a nontechnical audience. He pointed to the prevailing sentiment of public distrust of what they do not understand and of technology in general, and the pervasive public lack of confidence in government oversight. In his concluding remarks, Mr. Erwin cautioned that people may look to sources other than science for guidance in resolving troublesome technological issues when their culture is saturated with information and hyped with fear.

WORKSHOPS

In addition to attending the roundtable discussion and plenary sessions, NABC 5 participants each took part in one of four workshops: 1. Technical Risk Assessment and Regulations; 2. Public Assessments of Benefits and Risks; 3. Public Values: Benefits and Harms; and 4. Public Communication About Risk. The goal of the workshops, as at all NABC meetings, was to encourage frank discussion and to seek consensus, where possible, on the major issues in order to devise specific and useful recommendations for issue resolution and policy formation. While many of the stakeholders and interest groups represented at NABC 5 had attended previous NABC meetings, this was the first NABC meeting for the majority of attendees. There were an unusually large number of farmers and representatives of the print and electronic news media at the Purdue University meeting. For many of these people, this was also their first workshop experience.

Participants in each workshop spent six hours identifying and discussing key issues and developing recommendations of appropriate responses. Their recommendations were then presented for discussion by the entire group of participants during the final afternoon. A few highlights from the workshops are outlined below. For complete reports with a full list of recommendations, see Part II beginning on page 19.

Technical Risk Assessment and Regulations

Part of the public debate about agricultural biotechnology has focused on the measurement and regulation of risk. Discussion in this workshop cochaired by Rebecca Goldburg (Biologist, Environmental Defense Fund) and William Greenlee (Pharmacology and Toxicology, Purdue University), focused on aspects of risk characterization and risk management. Major issues identified by workshop participants included: 1. hazard identification of biotechnology products and processes; 2. establishment of scientific standards for measuring risks associated with biotechnology; and 3. better balancing of benefits and risks. Specific recommendations included:

> *More input is needed from the scientific community (e.g., commission study by the National Academy of Sciences) to develop hazard identification methodology for agricultural biotechnology products;*
>
> *Legislative gaps in regulatory authority should be filled (e.g., fish, shellfish);*
>
> *Land–grant universities need to address issues such as sustainable agriculture, family farms, pesticide use, for which biotechnology now serves as a lightening rod or even a surrogate focus;*
>
> *Models should be developed to assess the toxicity and allergenicity/antigenicity of expressed products as part of developing risk assessment guidelines.*

Public Assessments of Benefits and Risks

Public perceptions may make or break the introduction of agricultural biotechnology products. People's actions will decide which applications ultimately survive and which directions future research will take. Public perception research has shown diverse concerns about the risks associated with agricultural biotechnology. Participants in this workshop cochaired by Ted

McKinney (Community Affairs and Contributions, DowElanco) and Ann Sorensen (Center for Agriculture and the Environment, American Farmland Trust), discussed public reactions to agricultural biotechnology, societal ties between the people and their leaders, and the public policy formation process. Two major themes emerged from this workshop: 1. a possible paradigm shift in society's view of the role of humanity, science and technology in our world; and 2. the potential for biotechnology to significantly change our lives and our environment. The group's perception was that these themes are conflicting and suggest that the debate over acceptance of biotechnology will be influenced by more than scientific facts alone. Participants agreed on the need for better understanding of public values and attitudes toward biotechnology and its products. Specific recommendations from the workshop participants include the following:

Develop and implement methods of identifying and monitoring public understanding of and awareness about issues and potential changes being brought about through biotechnology;

Place additional emphasis within education and the educational process defining, assessing and understanding risk and decisionmaking under uncertainty;

Expand the capacity and commitment of the scientific community to more effectively communicate with the public;

Broaden involvement of stakeholders in identification of priority needs to be addressed by biotechnology.

Public Values: Benefits and Harms

Agricultural biotechnology has generated an ethical debate on actual and perceived benefits and harms. In some cases, scientists cannot measure risks because potential hazards have not been clearly delineated. Issues have arisen over the use of particular biotechnology processes (such as genetic engineering), the adoption of individual biotechnology products (such as genetically engineered tomatoes), and over decisions concerning regulation, research, intellectual property rights and other aspects of biotechnology. Each person views these issues through a unique framework of values composed of ethical, religious, economic, scientific and other beliefs. In the workshop on Public Values: Benefits and Harms cochaired by Rosetta Newsome (Scientific Affairs Information, Food Technologists) and Lilly–Marlene Russow (Philosophy, Purdue University), participants discussed why a diversity of views occurs, how these diverse views are formed and expressed, why or how they might be changed or harmonized, and who should make these decisions. The major issues identified by participants in this workshop included: 1. who should make decisions regarding biotechnology and its products; 2. what criteria are used to assign value to new biotechnology; 3. how safety is a factor affecting biotechnology; and 4. where communication about biotechnology needs to be improved. Following are some of the recommendations that were made to foster broader participation among stakeholders and to develop institutional mechanisms to enhance such an exchange of information and viewpoints:

Encourage and foster broad participation through a system responsive to stakeholders' input;

Be sensitive to religious concerns and provide information in food labeling accordingly;

Develop information which is clear and understandable, so as to be accessible to people with a variety of educational backgrounds;

Assess the social/economic impacts of specific biotechnology applications at the earliest state possible;

Single out and support land–grant universities and extension offices as particularly appropriate forums for discussion and dissemination of information;

Support better education at all levels, beginning with kindergarten.

Public Communication About Risk

The electronic and print media play a critical role in reporting information and ideas to the public. Editorials, newspaper and magazine articles, books, movies, and news and informational television programming all shape public understanding of and attitudes toward agricultural biotechnology. Effective risk communication requires that persons in the media have an understanding of the technology, the ability to use modern communication systems, and an awareness of people's concerns and interests. Participants in this workshop cochaired by Karen Bolluyt (Agricultural Information Services, Iowa State University) and David Judson (Gannett News Service) explored ways to conduct an effective dialogue on agricultural biotechnology to help society and consumers make informed long–term evaluations of the risks and benefits of biotechnology. Major issues identified included: 1. circular communication with effective consumer feedback must be developed and maintained; 2. improvements are needed to enhance the clarity and accuracy of the content of the message communicated to the public about biotechnology; and 3. greater effort is necessary to augment the credibility of biotechnology communicators. Specific recommendations include the following:

Focus on specific products or technologies, discussing risks or benefits for specific groups;

Base information on sound science, identifying the source's qualifications and affiliations;

Provide product information and access to process information;

Use language and concepts that the audience understands;

Share new and existing information from focus groups and surveys as widely as possible;

Employ mass media and targeted media to reach audiences, elicit responses from them, and build coalitions.

TYING IT TOGETHER

At the close of the conference, Dr. Theodore L. Hullar, Chancellor of the University of California, Davis and member of the NABC Council presented wrap–up comments for the group to consider as they returned to their daily professional routines. He noted how the meeting had captured society's frustration with the public policy issues surrounding biotechnology and had illuminated the centrality of social concerns regarding risk. Dr. Hullar commented on the uniqueness of the era in which discussions of biotechnology are held. He also explained the uniqueness of the issues raised by biotechnology. This uniqueness is derived from: 1. the scope and pervasiveness of the technology; 2. from the fact that biotechnology reaches closer to the centrality of life itself than do other technologies; and 3. agricultural biotechnology's enormous power because of its application to multiple potential targets, its dissemination into the environment, and its extension beyond the range of individual control.

In conclusion, Dr. Hullar affirmed his belief that the NABC format was working and that participants in this meeting would leave thinking differently as a direct result of the NABC experience. He challenged the participants to search for new risk evaluation paradigms and to engage in joint efforts, involving both social and biological scientists, to deal directly with the socioeconomic issues which biotechnology raises. Furthermore, he recommended that USDA proceed with a critical evaluation of socioeconomic issues associated with such topics as agricultural biotechnology and sustainable agriculture by fully funding basic and applied research on these topics through the National Research Initiative. He also supported the desirability of joint public–private partnership in the evaluation of the social implications of proposed new biotechnology products. Dr. Hullar concluded by emphasizing the need for these discussions to move from the realm of theory to the personal/practical level.

The fifth annual NABC open forum concluded with the new Chair of the NABC Council, Bill R. Baumgardt, Director of Purdue University's Agricultural Research Programs, charging the participants to take home with them the information from both the plenary sessions and workshops and from the many conversations in the corridors and over meals, and to implement the recommendations whenever and wherever possible. Reminding those present that individuals matter, he encouraged that broad, meaningful dialogue on agricultural biotechnology continue and wished all participants productive conversations about risk.

Putting It In Context

Theodore L. Hullar
Chancellor, University of California, Davis
(pictured on right)

Biotechnology[1] is, without doubt, one of the most precocious of discoveries, quickly moving into the center of biological sciences. It is providing for new products which promise to add much to commerce and industry, including specific biomedical therapies and dramatically improved agricultural practices and products. At the same time, biotechnology spawns strong, even fierce, controversy about manipulation of genome and environment alike. The annual meeting of the National Agricultural Biotechnology Council (NABC) addressed issues which confront our understanding of biotechnology, especially for agriculture. The purpose of this paper is to offer perspective on major features of this contentious terrain and to suggest some specific actions which might be usefully taken to clarify understanding and resolve issues.

THE CENTRALITY OF OUR SOCIAL ENVIRONMENT

Biotechnology engages the social environment, it would seem as fully as it engages the study of the investigator. The principle issue today with biotechnology is with the social environment within which it must necessarily function.

[1] Biotechnology, as used in this paper, is used generically to describe all manipulations at the molecular or cellular level that affect genetic material in a specific manner. Agricultural biotechnology, as used in this paper, refers to biotechnology on organisms and practices of importance to agriculture. It also refers, in the appropriate context, to biotechnology applied to food products and processes and also to environmental biotechnology.

We struggle to rationalize public concern about manipulating a single gene with the easy public acceptance of manipulating the whole plant or animal genome, such as it is, that has long been done in the traditional plant and animal breeding. We wonder how can it be that the possibilities of major manipulation by traditional breeding—where many genes are manipulated in unknown ways—are more socially acceptable than specific changes in a single gene or small set of related genes? Is it that we believe that an "undesirable" combination of genes in the traditional breeding will result in a lethal cross, and thus preclude socially "bad" crosses, whereas a nonlethal insertion of a single gene from a phylogenetically distant–related organism will give a nonlethal, but necessarily "bad" cross? Or is it something more?

Whatever it may be, the social issues of biotechnology were very early with us, starting with the Asilomar Conference in which scientists voluntarily agreed to monitor and control dissemination of their biological research, requiring their products and processes to pass voluntary government review, first with the Recombinant DNA Advisory Committee of the National Institutes of Health, and then for a counterpart committee for agricultural biotechnology managed by the U.S. Department of Agriculture (USDA). By all accounts, these methods are working very well. But, notwithstanding the obvious success, there is still broad social concern. And this concern leads to understandable frustration and steady struggle for understanding. It is likely true that these social concerns will continue to be with us until a body of experience and knowledge builds substantially further.

But I believe there is much more to these concerns.

BROADER DIMENSIONS OF THE CONTEXT FOR AGRICULTURAL BIOTECHNOLOGY

Biotechnology does not function in isolation. It is now an intimate part of our international technological fabric. It is part of our technological context, and it is thus afflicted by the same concerns besetting other technologies. Agricultural biotechnology, especially, has the potential to be pervasive through no effort once released into the environment. And the effects can be irreversible. It has the very real potential to create new life forms, possibly to obliterate old forms, and to spread its effects without control. In all of these ominous qualities it is similar to nuclear energy. More positively, it also has powers akin to those of the information revolution, the other molecular revolution in which we are engaged: powers to permit rapid, molecular–level changes; to give exceptionally high specificity and rapidity of effect; to tailor crops and agricultural practices in ways heretofore only dreamed about; to provide for major productivity increases, increasing the quality of produce in ways believed not

Biotechnology does not function in isolation.

possible. Because agricultural (and environmental) biotechnology is international, with the potential of affecting all manner of peoples and practices, it also becomes part of our cultural fabric. So biotechnology inextricably weaves itself through our technology and culture and like the information revolution, is seen as doing so unchecked by the normal forces guiding technological progress.

Agricultural biotechnology can, and likely soon will, focus on all the crops of the world; on most, if not all, of the cultural practices; and in virtually myriad environments. In all of this, it is much more complex and pervasive, and thus likely more vexatious, than biomedical biotechnology which focuses only on humankind, a single species with but little variation among its parts. This is a large part of our confounding context.

Biotechnology is also afflicted by the lack of trust in traditional leaders...

But there is more. Biotechnology gets to life itself. Its technological focus is control of genetic codes, of restructuring the genome, of making new forms of life, of tailoring life as we would have it be. Never before has there been this technological power, nor the number of expert practitioners working it.

It is no surprise, then, given the power of biotechnology over that which is most sacred—individual identity and character—that our skeptics are so troubled. They should be. Ours (scientists) is a trust that is, itself, sacred.

And this trust, itself, is wrapped in puzzlement. Discovery always is. What is being studied anyway? What value does it have? Who will control the results? And discovery in biotechnology, and the more difficult field of agricultural biotechnology, is even more vexatious because of the exceptional speed and specificity with which results can come. And control of the results is vested with the scientists themselves, or they are vested with the industrial laboratories that have, presumably, at least some self–interest in mind. So the cloak of silence of discovery roils the social context yet further. And this is exacerbated by the lack of personal control, perceived and real, over the results of biotechnology such as evidenced by concerns for the possible spread of genetic characteristics to unwanted organisms or to environments which are desired free of such interventions.

Biotechnology is also afflicted by the lack of trust in traditional leaders, such as scientists, government officials, major industries and university professors. We are perceived as out of touch, concerned with issues other than those of concern to society, unwilling or even unable to understand. Not much of that may be true, but it is widely believed, nonetheless.

Our agricultural biotechnology clearly has a complicated, intertwined, vexed context within which must be developed and used. And its human dimensions are especially important.

SEARCHING FOR PARADIGMS

Given the challenges, we must, foremost, be humbled by our knowledge and tools, and we must be awed by the responsibilities that are ours. And then we must necessarily consider carefully what to do to make progress, to create understanding, to make wise decisions, to be responsible stewards of that with which we have been entrusted. What are we to do?

First, we must never forget that individuals make a difference. Each has concerns and a life history that matter. We need to get information through the appropriate information channels to each of those who need and wish to know, and we must do so in ways each person can individually understand. Trust be engendered through straight talk, humility, and concern for truth and understanding.

Second, we need to understand what the relevant social structures are for agricultural biotechnology. Are they that of biomedical biotechnology? A patient–doctor relationship? An individual, willful undertaking? Or are they something different? I believe it much different because of the dramatic differences between biomedical biotechnology and agricultural biotechnology already referred to, such as pervasiveness in the environment, multiple and difficult–to–understand effects on many organisms, and evolutionarily permanent, at least in potential. This brings us directly to the importance of ecosystems and their functioning and stability, understandings we unfortunately know too little about, but which are so crucial to agricultural biotechnology.

Third, we need to do better at joining social science, values and ethics to our biological, physical and technological societies. Biotechnology proves to us that the separations between values and technology are nonexistent, or at least artificial. We had best fuse our concerns. The competitive grants program in the U.S. Department of Agriculture provides a new, significant avenue for making these connections.

Biotechnology proves to us that the separations between values and technology are nonexistent, or at least artificial.

Fourth, we must recognize that biotechnology is often a surrogate for other issues such as unchecked productivity increases, manipulation of the natural environment, changes in the structure of agriculture, continuation of technology–based agriculture as contrasted to some persons definition of sustainable agriculture, vertical integration and industrial hegemony in a heretofore highly decentralized and individualistic enterprise, unnatural means of producing food, and the like. What makes this so difficult is that each of these issues has plenty enough grist for the discussion and resolution mill without admixing them with biotechnology.

Fifth, to aid our understanding and decisionmaking we need to continue to work out the logical similarities and differences between analogies. Two such analogies are: 1. traditional plant and animal breeding (which could be called organismal biotechnology) as compared with biotechnology as now practiced (which could be called molecular and cellular biotechnology); and 2. biomedical biotechnology as compared to agricultural biotechnology.

...biotechnology is often a surrogate for other issues.

Sixth, progress in product development should be continued, with tough–minded but scientifically and procedurally fair reviews of products such as for approving genetically engineered fruits like the tomato and for changes in cultural practices as for herbicide–resistant plants and genetically engineered biological control agents.

Seventh, the NABC itself should continue to focus on being a crucible for testing similarities and differences of views, for finding common threads, for increasing understanding thereby.

Eighth, the USDA has some special opportunities and responsibilities which should be addressed forthwith: 1. The surrogate issues, outlined above, should be energetically and comprehensively dealt with, to the extent that is not already being done; 2. The relationships between agriculture and environment should be a special, ongoing emphasis. The two need not be in conflict. Indeed, they are not in inherent conflict. They are only made so by partisan adherents. Nowhere can this be more easily and productively addressed than in biotechnology for agricultural practice, concomitant with environmental improvement. A focused, integrated set of studies to this end—mutually undertaken by USDA, the Department of the Interior and the Environmental Protection Agency—should be established as soon as feasible, but not later than October 1, 1994; 3. The social science, ethics, and values issues embodied within agricultural biotechnology, as well as within the total agricultural enterprise, should be addressed through both basic and applied research convoked through the Department's competitive grants program; 4. evaluation protocols appropriate for agricultural and food (and environmental)biotechnologies should be developed distinct from those used for biomedical biotechnology, as has been oft–noted at this conference. This study could effectively be done by the National Research Council through its Board on Agriculture in collaboration with the Council's Food and Nutrition Board and its Commission on Life Sciences.

Lastly, these issues can, and must, be considered at the intellectual, even abstruse level, which university faculty enjoy. But, the issues are real and they are ultimately felt by all humankind the world over. Our challenge, then, must be to deal with theory and rational analysis, as is our wont, but we must also be sure we deal, ultimately, with the issues in the fundamentally human and individual terms that are, after all, the real focus of our attention.

PART II

Workshop Reports

HOW TO MEASURE RISKS

Technical Risk Assessment and Regulations

cochairs: Rebecca Goldburg, Biologist, Environmental Defense Fund
William F. Greenlee, Pharmacology and Toxicology, Purdue University

Products of agricultural biotechnology, such as field tests of genetically engineered crops or foods derived from genetically engineered crops, may pose risks to ecosystems or human health. However, the traditional risk assessment paradigm, developed to assess the carcinogenicity of chemicals, is not easily applied to products of agricultural biotechnology. Thus it is necessary to develop new risk assessment approaches in order to assess the risks of many agricultural biotechnology products.

After making scientific assessments of the nature and magnitude of any risks, regulators and other decisionmakers must elect a course of action. This risk management process often involves weighing risks and benefits of a particular product. The process can be difficult for agricultural products (e.g., pesticides, whether genetically engineered or not), since many of the individuals who bear direct risks from these products may not be the primary beneficiaries of the products.

Workshop participants set out to identify issues and make recommendations concerning risk characterization and risk management in agricultural biotechnology. Participants were first split into three groups to identify important issues. From the large number of issues identified by all three groups, participants selected, by vote, three issues for further discussion. The selected issues were essentially consecutive steps in the risk characterization and management process:

Identify hazards of process/product
Measure risks and establish scientific standards
Balance risk and benefits

Workshop participants then divided back into three groups, one for each issue, to develop recommendations. These three groups reported their recommendations to the workshop as a whole, and all participants were given an opportunity to discuss the recommendations before they were made final. The third group noted that risk is a part of life, and many participants felt that consideration of biotechnology products should somehow involve benefits as well as risks. Balancing risks and benefits for agricultural biotechnology products can be extremely difficult, however, especially when risks and benefits are not, for the most part, borne by the same individuals or groups. No specific recommendations were agreed upon for balancing risks and benefits.

RECOMMENDATIONS

IDENTIFY HAZARDS OF PROCESS/PRODUCT

The National Academy of Sciences should study and develop strategies for hazard identification in agricultural biotechnology.

Regulators have extensive experience identifying the hazards of synthetic chemicals, but this experience is not always directly applicable to agricultural biotechnology products.

More input is needed from the scientific community to develop hazard identification methodology for agricultural biotechnology products.

Legislative gaps should be filled (e.g., fish, shellfish).

Regulatory agencies, in some instances, lack the authority to adequately address risks of agricultural biotechnology products. Fish and shellfish present a clear example of such a gap. No agency has a clear Congressional mandate to regulate either the risks of releases of genetically engineered fish and shellfish or to regulate the safety of fish and shellfish (genetically engineered or not) for human consumption.

Land-grant universities need to address issues such as sustainable agriculture, family farms, and pesticide use, for which biotechnology now serves as a lightening rod or even a surrogate focus.

Some agricultural biotechnology products are the focus of considerable criticism or opposition from individuals who believe that these products may exacerbate existing trends in agriculture. Many issues about the structure of agriculture, (e.g., the loss of family farms), merit public debate. Unfortunately, few obvious forums are now available for public discussion of these issues. As a result, in some cases biotechnology products are serving as the primary vehicle for debate.

Government officials need to develop integrated approaches to regulation that incorporate knowledge of product and process.

As has been noted by many others, risk assessment of biotechnology products should be based on the characteristics of the products and not the fact that biotechnology was used to develop the product. Nevertheless, knowledge of the process used to develop a product can sometimes help form the questions asked in risk assessment or aid in decisions concerning which products to assess (e.g., in assessing the safety of a drug, regulators often consider the process used in its manufacture because the process can affect the presence of impurities in the product). Regulation of biotechnology products should be based on science and the law, and not on ideological avoidance of all references to biotechnology.

Measure Risks and Establish Scientific Standards

Tools (appropriate test systems) should be developed to evaluate the potential hazards of three classes of organisms: animals, plants and microorganisms.

Scientific guidelines need to be developed to ensure that any ecological or human and animal health risks of agricultural biotechnology products are adequately addressed. The following are examples of areas that may merit the development of such "tools:"

Animals
- –containment/ecological effect of releases
- –human safety of expressed products
- –unforeseen metabolic effects

Plants
- –containment/ecological effect of releases:
  - altered disease/insect susceptibility,
  - weediness, and
  - outcrossing.
- –human safety of expressed products
- –unforeseen metabolic effects

Microorganisms
- –containment/ecological effect of releases:
  - colonization,
  - pathogenicity/toxicity to nontarget organisms, and
  - frequency and impact of gene transfer to other microbial species.
- –unforeseen metabolic effects

Models should be developed to assess the toxicity and allergenicity/antigenicity of expressed products as part of developing risk assessment guidelines.

It also should be noted that it is impracticable to measure any and all potential unforeseen effects. One can only look for specified unforeseen effects of particular concern.

methods of
understanding and
changes being
Place additional emphasis on
K-12.

Public Assessments of Benefits and Risks

cochairs: Ted A. McKinney, Community Affairs & Contributions, DowElanco
A. Ann Sorenson, Center for Agriculture and the Environment, American Farmland Trust
with Patrick A. Stewart, American Farmland Trust

The purpose of this workshop was to establish an understanding of the underlying reasons for public concerns about agricultural biotechnology. This was accomplished by dividing the 28 participants into three smaller groups to discuss issues and possible solutions followed by an open group discussion in order to reach consensus. Over the course of two days, the often spirited discussion revolved around public perceptions and ways in which to respond to the public's need for credible information.

MAJOR THEMES

Discussions centered around two interrelated themes that came into sharp focus later on in the workshop. Both dealt with public perceptions of biotechnology. The first was a possible paradigm shift in the way the public thinks about the benefits and risks of biotechnology. The perception of the group was that there is no longer an unquestioning acceptance of the social paradigm in which humans are seen as dominating the planet and its resources. It is being replaced by an environmental paradigm, in which the public perceives limits to growth. Biotechnology's place in this new paradigm has not yet been determined, but its perceived role may affect its acceptance.

The second underlying theme that emerged from the workshop was that of the potential of biotechnology to bring about change. With the introduction and development of biotechnology, society has been given a powerful tool to change its environment in unforeseen ways. This leads to questions about how society should shape itself and who should make decisions as to the form and extent of change. This has understandably raised public concerns. However, the workshop participants saw that decisions pertaining to biotechnology must be made because the technology cannot be suppressed, only channeled into desired uses. The participants further saw that if the public did not become involved in the decisionmaking process early on, the marketplace would make decisions in its absence.

These two themes are seemingly in conflict. On the one hand, many people are fearful of new technologies and express an increasing desire for nonintervention in the ecology of the planet. On the other hand, biotechnology offers the possibility of more controlled and targeted interventions into the environment, minimizing negative side–effects. Ironically, even as science has developed a technology that can change the environment, the public's perception and acceptance of biotechnology has turned away from the promise of benefits, and focused on risks associated with technological change. Technical analysis of benefits and risks is no longer sufficient to assure acceptance of biotechnology. Increasingly, an appraisal of the public's perception of those risks and benefits by stakeholders must be considered as well.

STAKEHOLDERS

The participants felt it was important to define who the stakeholders in public assessment of risks and benefits associated with agricultural biotechnology are. The stakeholders fall into five broad groupings:

Government: Local, state and federal level policymakers in the legislative, executive/administrative and judicial branches.

Universities/Research Organizations: Groups providing the scientific knowledge and information on which policy decisions are often based.

Special Interest Groups/Organizations: Groups with an interest in preserving or changing the social and economic status quo. These include environmental groups, farm groups, industry/trade organizations, health groups, unions, religious groups and others.

Corporations/Organizations Funding Research: Organizations creating products and technologies to serve their constituents/consumers.

Consumers/The Public: Individuals directly affected by the production and consumption of biotechnology.

ISSUES

Workshop participants identified three major issue areas in public perception of agricultural biotechnology: personal issues, societal issues and process issues. Intervention to influence public policy can take place at any of these levels.

Personal Issues

Personal issues deal with how an individual views the impact of biotechnology on themselves. Personal issues identified by workshop participants included: food safety/health, economic impact, environmental/animal health, and the spiritual/moral–ethical dimensions of genetic engineering.

Food Safety/Health: Questions about the safety of food manipulated through genetic engineering and its impact on human health are of particular concern for the agricultural biotechnology industry.

Economic Impact: Individual concerns over the economic impact of agricultural biotechnology center around possible positive and negative impacts on jobs and the introduction of new products into the market.

Environmental/Animal Health: Individuals are becoming increasingly aware of negative impacts to both the environment and animals. The perception of agricultural biotechnology's role in ameliorating or aggravating current conditions will influence the acceptance and use of agricultural biotechnology.

Spiritual/Moral–Ethical Dimensions: Some individuals seem to have a "gut reaction" to biotechnology, not necessarily connected with religious beliefs, that no amount of scientific knowledge or data will change.

Societal Issues

Societal issues deal with ties between society and its leaders, and the strength of those ties. Again, in no particular order, the participants identified the following issues:

Trust: Public involvement in policymaking has increased in the past two decades. Citizens are increasingly suspicious that policymakers, who they see as manipulated by private interests, may be trying to manipulate them. This leaves them wondering who to trust with decisionmaking and who to turn to for credible information. Debate over the scientific accuracy of information also has contributed to a basic distrust of authority.

Motives: The public also has developed a distrust of some individuals and groups on the basis of their motives. They perceive an imbalance between those who bear the risk and those who benefit in society. The brunt of this distrust has been directed at industry because of its necessary focus on profit–making.

Socioeconomic Concerns: There was concern over the possible impacts of biotechnology on the socioeconomic structure of groups and communities. The possibility of concentration of ownership of food production through control of biotechnology is disconcerting to some. The potential impact of agricultural biotechnology on small and medium–size farms, and its possible contribution to the loss of a traditional way of life, is of special concern.

Process Issues

The third tier of concern was that of process issues, or how policies are made and who makes them.

Public Policy: Questions over public policy on biotechnology focused on who determines policy, who should determine the policy, and how policy should be implemented. There was an expressed need to find a process of technology control with which the public is comfortable and involved.

Public Understanding of Agricultural Biotechnology: The issue of public knowledge was seen as going beyond that of agricultural biotechnology to science and technology as a whole. Participants felt that lack of public knowledge was understated by most people and that some people simply are not concerned or likely to ever be concerned about the impacts of biotechnology on their lives. In spite of this, there was a desire to distribute accurate information ranging from the technical aspects, the socioeconomic impacts and the moral/ethical concerns of biotechnology.

Public Attitude About Biotechnology: There was a two–pronged question about the public's attitude concerning agricultural biotechnology and how to assess it. The public currently receives information from competing channels: special interest groups, media, government, industry and university sources. Because these messages often conflict, public attitudes about biotechnology are shaped by what appears to be the most trustworthy source. It is at this point that information moves from being a scientific issue to being a political issue.

Process and Value of Measuring Public Response to Risk: The value of measuring public perception of risk was questioned for two reasons. First, gaps between the public's statements of what it would do and what it actually does always exist. Second, even if such information were known, changing public attitudes is difficult.

What Does the Public Think vs. What is Known: There was a belief that the majority of public decisions are made on the basis of emotion and limited information. This is of great concern to researchers and industry who must deal with a public that cares little about science and must attempt to bridge the gap between science and the public's understanding of it.

SUMMARY STATEMENT

After much discussion, workshop participants agreed on the following summary statement on public assessment of risks and benefits of biotechnology:

> We recognize that technical assessment is not the only factor in public acceptance of technology. We recognize the need for better understanding of personal and societal values. We also recognize the need to understand the factors influencing public attitudes about biotechnology and biotechnology–derived products on the part of stakeholders.

RECOMMENDATIONS

On the basis of the preliminary discussions, the combined group developed recommendations for dealing with personal, societal and process issues. These objectives address the concerns outlined in the group's summary statement:

Develop and implement methods of identifying and monitoring public understanding of and awareness about issues and potential changes being brought about through biotechnology.

Greater support is needed for social science research through multiple methods such as surveys, informal information gathering, expanded dialogues between stakeholders, public forums and content analysis.

Place additional emphasis within education and the educational process on defining, assessing, and understanding risk and decisionmaking under uncertainty.

Expand the capacity and commitment of the scientific community to more effectively communicate with the public.

Expand to an ongoing dialogue about the implications of the knowledge being generated.

Place additional emphasis on science education in kindergarten through twelfth grade.

Expand public dialogue and discussion about the forces of change being generated by biotechnological developments beyond traditional channels (the Federal Register, Public Comment, *university extension services, etc.) in order to reach the public at the grassroots level.*

Broaden involvement of stakeholders in identification of priority needs to be addressed by biotechnology.

Public Values: Benefits and Harms

cochairs: Rosetta Newsome, Scientific Affairs and Information, Institute for Food Technologists
Lilly-Marlene Russow, Philosophy, Purdue University

The workshop began with clarification of the topic of the workshop, and how it might be separated from the other topics. It was noted that the concepts of benefit and harm—particularly the latter—were considerably broader than the more specific idea of risk. "Risk" tends to invite a focus on health and safety issues, while benefits and harms extend beyond these specific concerns. Nonetheless, it was clear that the topic is very broad, that nearly every question about biotechnology is a question about public values, and hence that any attempt to predetermine the focus of the workshop would restrict the discussion too much. Two themes were repeatedly emphasized: 1. public values grow out of attempts to acknowledge and balance the values of diverse individuals; and 2. the whole issue of public values must consider the process by which values are shaped, expressed and recognized.

IDENTIFYING ISSUES

The participants were asked first to identify, and then to prioritize, topics of greatest concern. The results were as follows. Issues are listed in the order they were assigned as the result of vote, with comments voiced by participants and examples of topics within specific issues included under each general heading[1]. Topics are reproduced exactly as formulated by the workshop, since in many cases there was substantial debate about the wording. The results of the voting upon which the ranking was based are included. The first issue, identified below, ranked considerably higher than the others[2].

[1] The summaries under the heading "Elaboration/Analysis of Major Issues" reflect only the initial discussion. The workshop discussed the top–ranked issues in more depth on the second day; the discussions are summarized under the heading "Other Issues."

[2] Each participant was asked to list all seven topics in order of importance. Each list was then weighed, with the issue listed first receiving a "1", the second a "2", and so on to the last, which received a "7." Since there were sixteen participants who voted, there was a possible range of 16–112 points, with the lower numbers representing the issues judged more important.

Who should have the right or power to make decisions that have broad social implications? (29 points). Participants decided to consider the question of who ought to have a voice, rather than simply to ask how decisions are currently made. Participants pointed out that public institutions are poorly funded and need to be empowered, and that a sharp, reductionistic divide between science and technology and other sources of value (e.g., religion or spirituality) exists. More generally, this topic encompasses the questions of who sets research agendas, who shapes and controls the regulatory process, how the food–production system is determined and controlled, and who decides what products are available.

What criteria are used to assign value to new biotechnology? (53 points). The discussion began with a look at how the public views biotechnology and other "new" technologies in contrast with familiar products and processes. Some view new scientific discoveries as "progress," and "new" as equivalent to "better" or "improved." However, at least since Hiroshima, others express increasing numbers of questions, and perhaps skepticism, about the wisdom and value of some so–called "advances."

How safe is safe enough? (59 points). This discussion began with the observation that people today have different expectations about safety than they did earlier; they are more likely to raise questions about the safety of everything from food to playgrounds than they were fifty years ago. Although safety is only one factor in public value, it is important enough, and complex enough, to warrant careful consideration. Participants noted that there were conflicting ideas about what is included in judgments about safety. Perceptions of safety were tied to control in that something one can choose to avoid (e.g., bungee jumping) is less likely to raise serious safety concerns than things that are more difficult to avoid (e.g., drinking water). It was also pointed out that there are discrepancies between what people say they want and what they are willing to pay for, but that economic and class value systems were important factors to keep in mind. The importance of avoiding an elitist structure was emphasized: safety should not be a luxury limited to those who can afford to pay for it. On the other hand, concern about safety rarely overrides basic needs—one participant mentioned that people starving in Sudan, or even getting canned food from a soup kitchen, are less likely to worry about insect damage or contamination than affluent Americans.

What communication is needed among all citizens affected by biotechnology? (65 points). Preliminary versions of this topic were phrased in terms of information, but people soon changed to a discussion of communication in order to emphasize the need for true dialogue and the importance of avoiding an arrogant "us vs. them" attitude. Specific questions noted under this general heading were: 1. what sort of information consumers need and want; 2. the concept of "informed consent" and what that standard requires;

and 3. what sort of information regulators and legislators need in order to reach decisions.

How does biotechnology affect distribution of assets, incomes and power? (73 points). In general, public values will vary according to who wins and who loses through biotechnological advances. If large corporations are perceived as profiting while small family farms are perceived as harmed (as the bST/bGH controversy is sometimes portrayed), biotechnology is more likely to be viewed as harm. A more specific subheading under this issue had to do with the impact of biotechnology on the structure of the food production and distribution system.

What is the environmental impact of biotechnology? (85 points). There was little initial discussion of this issue, but subsequent comments indicated that it included, among other things, affect on the type and quantity of pesticide used, water quality and biodiversity. On the issue of biodiversity, one participant pointed out that more thought must be given to the choice of plant species used, e.g., to develop substances such as plastic substitutes. Choosing alfalfa rather than corn as a 'host' for example, the participant said, would be beneficial in that it would help stem the tendency towards monocultures and their attendant problems.

How should concerns for animals be taken into account? (93 points). There was considerable debate about the wording of this point. Some people wanted to describe the issue in terms of a contrast between concern for animal welfare and animal rights ("Should animal welfare be expanded to include animal rights?"). Others felt that terms like "animal rights" were prejudicial and unclear, and that a broader and more neutral description of the issue would be preferable. By majority vote, the form given above was chosen by the group.

ELABORATION/ANALYSIS OF MAJOR ISSUES

The workshop was then asked to break into two subgroups to explore the two issues identified as the most important topics. Each group was given one topic, and asked to identify: 1. barriers which hindered the group from addressing the issue effectively and appropriately, and; 2. recommendations about how to deal with the issue and the associated barriers. The barriers and recommendations developed by each subgroup were then presented to the entire workshop for discussion.

Who Should Have The Right or Power To Make Decisions That Have Broad Social Implications?

The subgroup which discussed this highest–ranking issue began by identifying the various sorts of "players" in the process of evaluating biotechnology. The following were identified: 1. regulators (including legislators and regulatory agencies such as USDA, FDA, etc.); 2. developers (industry, scientists, granting agencies, academia); and 3. consumers (including both consumers in the literal sense of people who buy a product and also people seeking other benefits, such as environmental groups). The news media were also cited as players which can wield significant influence.

Next, the group identified the following barriers:

—Exactly what people want to know is not always known.

—Not everyone wants to get involved; some people want someone else to make the decisions. Consumers typically are overwhelmed; scientists typically want to be left alone to focus on their own work.

—Current procedures for gathering views and disseminating information are too formal to be widely effective (e.g., most people do not read the *Federal Register*).

—A common base of shared knowledge cannot be presupposed.

—In determining value, the scientific processes which are learned, the investigative tools which are developed and the advances in basic science, not just the concrete products of biotechnology need to be considered.

—The complexities of diverse cultures and value systems need to be understood and respected. This will affect, among other things, choices about whether, when and how, to compete with other countries in the international marketplace.

—Although the current political forum in which policy is shaped is supposed to be democratic, questions are raised concerning how democratic it is in practice.

Recommendations

Finally, the group offered two recommendations:

Increase the opportunity for "friendly" participation in the formal process. A system is needed which encourages and fosters broad participation, and which really listens and responds to input from all stakeholders.

Congressional hearings, by contrast, are often unfriendly, and people who testify often leave with the feeling that their input made no difference.

Real discussion (as opposed to mere dissemination of information) needs to be promoted among broad and diverse audiences.

NABC meetings represent a valuable first step, but do not represent the diversity of positions and values that must ultimately be included.

What Criteria Are Used To Assign Value To New Biotechnology?

The second subgroup identified barriers and made recommendations regarding this second major issue. Barriers fell into two major categories. Several examples are identified in each category.

The first category of barriers or complexities in defining value or public good concerned the heterogeneity of people affected by biotechnology and public values. Religious, ethnic, economic, age and educational differences were all thought to effect how people judge the benefits and harms of biotechnology. Specific biotechnological developments will rarely be perceived similarly by all segments of the population. Moreover, people differ with respect to their willingness and/or ability to accept risks.

The second category of barriers related to information, and the difficulty of getting information into a public forum early enough. Full and free exchange of information is often hindered by concerns for intellectual property rights, the proprietary interests of an industry and competitiveness between industries, the desire of scientists to keep findings to themselves until their work has been published, and regulatory restrictions on discussion of products under regulatory review. It was suggested that there is a possible "window of opportunity" for earlier exchange of information after a patent has been granted, yet prior to marketing. This suggestion was countered by the observation that the restrictions of the patent process limit this potential "window." Participants recognized that within the current system, a significant investment, both public and private, is made before public value is fully established.

Recommendations

The group then offered several specific recommendations to help overcome the problems inherent in dealing with the heterogenicity of the public. While the general theme reflected an encouragement of broad public involvement and consideration, these items were considered more as examples than as a complete list. The following suggestions were identified:

Be sensitive to religious concerns and provide information in food labeling accordingly.

Develop information which is clear and understandable, so as to be accessible to people with a variety of educational backgrounds.

Assess the social/economic impacts of specific biotechnology applications at the earliest stage possible.

Specifically, applications should not adversely impact individuals in the low income sector, e.g., applications that would raise significantly the cost of foods should be avoided.

Establish a societal "minimum acceptable risk level," recognizing that some products or processes might be too risky to be acceptable at all; and identify risk levels of acceptable applications to enable individuals to make personal decisions about risk acceptability.

To deal with information barriers, the group suggested:

Land–grant universities and extension offices be singled out as particularly appropriate forums for discussion and dissemination of information.
However, effective functioning in these roles requires increased funding, more attention to and respect for extension activities as part of the original mission of land–grant institutions, and more autonomy from industry support.

More attention be given to the "window of opportunity" (see above).
Information should be exchanged and made available as widely as possible during this period, and consideration should be given to modifying the processes and regulations to allow for better exchange of information as soon as possible. Thus, open forums (designed to encourage personal communication and dialogue, not promotion) during the early stages of development before beginning marketing, need to be fostered. This would require, among other things, clarification of restrictions on discussion of patent applications under review and products under regulatory review.

Finally, participants offered a variety of additional criteria likely to arise in various applications of biotechnology. This list is not to be interpreted as recommended standards or criteria to be formally incorporated into the approval process, but rather, items which warrant consideration as early as possible in the developmental process. The difficulty of accurately projecting impact of various applications was recognized, though. The first point mentioned in this regard was the need to ***pay attention to both long–term and short–term impact***: people evaluate a product on the basis of what its impact may be in twenty years as well as what it may be now.

Other criteria mentioned were: ***impact on the food supply***—nutritional value, food quantity, quality, variety and cost, and ***impact on the structure of agriculture.*** With regard to agricultural structure, concerns were: Will the application accelerate vertical integration and the role of farmers in the decisionmaking process and will it impact the sustainability of the process? Is government support required? Impact of the new development on current products and on food production were also offered.

Environmental concerns generated another set of possible criteria. Water quality, sustainablilty and biodiversity were a few of the concerns that were highlighted. ***International effects*** also were mentioned, including competitiveness and impact on Third World or developing nations. Finally, ***economic and social considerations*** were brought up again, with particular emphasis on concern for equitable distribution of financial gains.

OTHER ISSUES

The detailed discussion of the two most important issues did not leave much time to pursue the other topics that had been identified during the first session. Nonetheless, some of the lower–ranking topics (e.g., the environment) had been taken up in the course of discussing one or both of the first two issues, and some discussion of the third and fourth issues (safety and communication) was possible. The variety of participant comments on the third and fourth issues are detailed below.

How Safe Is Safe Enough?

Several barriers were noted. The first was short–term versus long–term safety considerations. A related point emphasized generational considerations, and the fact that people are often willing to take risks for themselves, but not for their children. (Alar was cited as an example of this point). Second, the public is increasingly unwilling to trust science and industry, and to view assurances from these sectors with suspicion. Next, the tension that often arises between public safety and individual freedom and choice was mentioned. Laws requiring motorcycle helmets, and New Jersey's short–lived attempt to prevent restaurants from serving soft–boiled or sunny–side up eggs were cited as examples. Finally, the apparent failure of our educational system to provide people with an adequate understanding of scientific methods and the limits of science, for example, was cited.

Recommendations

These problems gave rise several recommendations:

The need for better education at all levels, beginning with kindergarten.

The need to avoid absolutes when talking about safety.

Nothing is simply or absolutely safe, and this requires open communication about levels of safety.

It is necessary to take a much broader perspective when considering safety.

One should attempt to evaluate the whole process, source as well as outcome. In considering *Salmonella* contamination for example, all stages of the poultry and egg production process are to be evaluated, not just egg preparation and consumption practices.

What Communication Is Needed Among All Citizens Affected By Biotechnology?

This last issue addressed the need to improve communication. The main point conveyed here was the need to communicate on an effective, personal level, which requires, among other things, listening to public concerns as well as providing information in an appropriate way. Educational levels and vocabulary were two factors that should be kept in mind. Realism is always necessary; practitioners must listen to real situations expressed by the public. It also was recognized that many individuals exhibit a narrow vision or focus on their own specific agenda, and that communication may be hampered by people's unwillingness to get involved, or doubt about whether they should really care about these issues.

Recommendations

To address these problems, the group recommended:

Better support for land–grant institutions and extension offices.

Better education programs at the K–12 levels.

More strenuous efforts to support scientific societies (e.g., Institute of Food Technologists) with information for broad dissemination.

Public Communication About Risk

cochairs: Karen Bolluyt, Agricultural Information Services, Iowa State University
David Judson, Gannett News Service

MISSION

To provide society the information it needs to evaluate the potential risks and benefits of agricultural biotechnology.

BACKGROUND

Certain basic principles can guide all involved in public communication about risk as it relates to biotechnology.

Opinions about risk vary from one perceived risk to another. Peter Sandman, professor of environmental communication at Rutgers University, has described "outrage factors" that drive personal assessments of risk. Communication about any perceived risk should include an analysis of the risk in light of factors that influence personal perceptions of risk. According to Sandman, these include: 1. individual control in assuming risk (voluntary vs. involuntary exposure); 2. fairness or the extent to which a risk is distributed equally; 3. morality or the extent to which technology or behavior not only poses a risk but is perceived to be evil; 4. dread, e.g., the belief that the potential damage may be catastrophic or may cause a fatal, lingering illness; 5. familiarity, as illustrated in the difference between fear that peanut butter may contain carcinogens or the fear that irradiation may change foods in undesirable ways; and 6. trust as earned or lost in all areas of organizational behavior.

Some elements of modern technology help cause increased perceptions of risk. These include: 1. the improved ability to detect toxic substances (one part per quintillion); 2. new technology that is not understood except by people with exceptional skills or highly specialized education; 3. knowledge of catastrophes or instances in which technology believed to be beneficial proved to be harmful (e.g., thalidomide); 4. experts disagreeing during litigation, hearings or other widely publicized public discussions; 5. growing production and distribution systems that increase the potential for technologies and products to affect millions of people each day, thus increasing the chances for catastrophe; and 6. growth in knowledge and the accompanying growth in awareness of gaps in knowledge (How valid are methods of risk assessment?).

Categorizing some perceptions of risk as "irrational fears" interferes with risk communication and is a counterproductive substitution for thoughtful exploration of issues/answers. "Many risk experts insist that 'the data' alone, not the 'irrational' public, should determine policy. When a risk manager continues to ignore [outrage] factors—and continues to be surprised by the public's response of outrage—it is worth asking just whose behavior is irrational." (Sandman, 1987)

The long–term view for risk communication is that society and/or consumers determine the success or failure of new technology and new products. This long–term outlook should drive communications plans and activities.

In the United States a majority of people express some belief that biotechnology in agriculture can benefit them and express some support for the development of biotechnology. The public strongly expresses a need/desire for information about biotechnology and for the opportunity to be involved in decisions about the use of biotechnology in the development and use of products.

Communication is not the easy task of message distribution once the difficult decisions about financing, research, development, marketing, etc. are made. It is a crucial, complex, continuous, circular interchange that should be a central part of all planning and budgeting. In general, communication plans and efforts have been inadequate.

RECOMMENDATIONS

The workshop participants prepared recommendations on three topics: communication content, credible communication, and circular communication. Some recommendations were made for more than one topic, but each is reported only once here.

Communication Content

Communication should contain more than facts. Opinions and values should enter the communication mix at every juncture, and that is taken into account later in this report. The following recommendations regarding factual information were made:

Communicate in specifics as much as possible.

Focus on specific products or technologies, risks or benefits.

Focus on what a product/technology will mean to specific audiences.
Use simple language (old, short words).

Prepare to be brief and concise about key ideas and information, and be prepared to provide detail (probably written).

For all sources of information, identify the source's qualifications and affiliations.

Base information on sound science.

Credible Communication

Beginning with the assumption that credibility must be earned and conferred, that it can not be claimed or bought, the following recommendations were made:

Provide full disclosure of information about benefits, risks, and the assumptions on which the information is based.

Be clear and forthright in describing biases or financial interests that an audience should understand to evaluate information and opinions from various sources.

Provide product information and 1. provide, or 2. make it easy to obtain process information.

Do not simply state conclusions; provide background information about how conclusions were reached and distinguish between opinion and fact.

Use language and concepts that the audience understands.

Clarity is credible. People are more likely to be suspicious of what they do not understand.

Choose spokespeople carefully, considering each audience and using the audience's criteria for trustworthiness.

Build bridges with key groups by identifying people who can serve as liaisons.

All members of most groups will not become experts in biotechnology, so they identify a trusted group member or liaison who is knowledgeable and they rely on him or her for guidance.

Circular Communication

If one accepts the proposition that the consumer will be one of the primary determining factors in the process of acceptance, then there are two critical communication questions to be addressed: 1. How do we provide the information consumers need? and 2. How do we develop and maintain effective feedback from consumers? This process is complex, but it has a circular nature that provides points of reference for plans and actions. These points of reference are: 1. provide information for the forum of pubic debate; 2. listen to the feedback in the ensuing dialogue; and 3. go back to point one. The following recommendations for establishing circular communication are made:

Listen more than you talk.

All participants should make special efforts to listen attentively, with the goal of understanding the facts and beliefs behind various points of view. A corollary to this is that multiple sides of an issue should be presented during meetings/discussions.

Adopt the wheel model for risk communication. (right)

The hub is society/the consumer. The spokes are conduits for information flowing to and from the public, and the rim is the arena of interaction containing multiple sources of information and opinion. Any poorly functioning part of the wheel will have a negative impact on the whole.

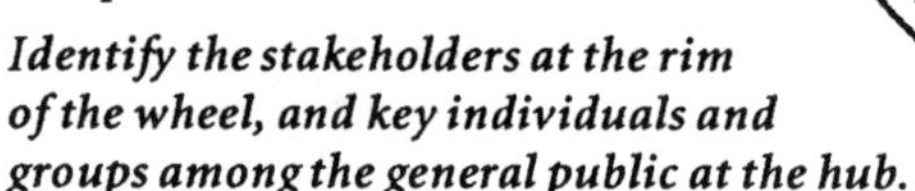

Identify the stakeholders at the rim of the wheel, and key individuals and groups among the general public at the hub.

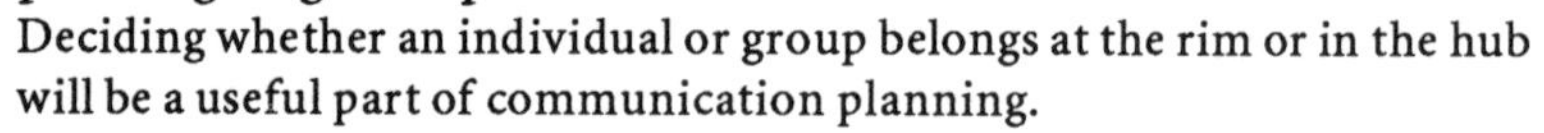

Deciding whether an individual or group belongs at the rim or in the hub will be a useful part of communication planning.

Identify the communication channels.

Define the role, the costs and the importance of each channel, and develop communication plans accordingly. Among the channels that might be used to exchange information are the following: K–12 education, land–grant institutions—particularly their extension services, interpersonal communication, mass media, targeted media, coalitions, focus groups and surveys, consumer behavior, organizational boards, and formal or informal opinion leaders.

The following were selected recommendations for channels of communication:

Increase funding for such programs as "Ag in the Classroom."

Science teachers' associations should be invited to cooperate in planning educational programs.

Provide scholarships for teachers and students that could bring them to university and industry labs as interns or workshop participants.

Take advantage of all opportunities to build coalitions. Bringing together groups that disagree often works.

Areas of disagreement are based partly in misunderstandings and lack of information. Common ground and common goals often can be identified. Such coalitions can become credible communications channels because they do not represent a single point of view.

Share new and existing information from focus groups and surveys as widely as possible.

This is one efficient way to identify problems and issues early and to build general understanding of biotechnology and of public opinion and behavior.

Use the land–grant model for coordinating communication.

Cooperative Extension might be empowered to build and coordinate cooperative agricultural biotechnology communication programs.

Identify and use opportunities for interpersonal communication.

Several decades ago, research on the adoption of innovations pointed to the importance of interpersonal communications for decisionmaking. Recent research on risk communication indicates that human behavior has not changed in this regard. This inefficient channel for communication may be the only effective channel/best channel in many instances. Organizations should make interpersonal communication (i.e., listening and talking) a strategic part of communication plans.

Use mass media and targeted media to reach audiences and to elicit responses from them.

Media relations strategies have changed considerably with the growth of special interest publications. There never has been any such creature as "the general public," but media targeted at specific groups have increased in importance while many "mass media," (e.g., daily newspapers) have decreased. The role of mass media and more targeted media as sources of facts and as mechanisms for calling attention to issues also has remained relatively stable since the time of the adoption–diffusion studies of the 1940s.

Use the body of communications research on the role of these communications channels to plan risk communications.

Advise organizations and institutions to incorporate diverse points of view into their leadership.

This should begin with The National Agricultural Biotechnology Council.

Recognize the importance of informal and formal leaders.

For many issues, formal and informal leaders are sought out for their opinions. Sometimes they are in decisionmaking positions, but not always. They are, however, channels for information. Special efforts should be made to understand how they obtain information and to keep them informed.

Communication about biotechnology is a complex process that requires equal attention to facts about the science and understanding of human behavior. It requires planning, resources and respect for the consumer. It can be frustrating. Poorly executed, it can create ill will and a great drain on resources directed at damage control. It should receive careful as the attention from the beginning of any efforts in biotechnology.

REFERENCES

Sandman, P. 1987. Risk Communication: Facing Public Outrage. *EPA Journal.* 13:21–22.

Plenary Lectures

*Regulatory Risk Assessment: A View from the Potomac**

David R. MacKenzie
Director, National Biological Impact Assessment Program USDA;
Principal Plant Scientist, Plant and Animal Sciences CSRS–USDA

One needs to understand the history of U.S. biotechnology regulations in order to comprehend the present regulatory structure. It all began with the recombinant DNA research in the early 1970s. At that time the hazards of the research were not known and the scientific community formulated its own program for biosafety oversight, managed through the National Institutes of Health (NIH), coordinated through the Recombinant DNA Advisory Committee (RAC), and operated through a distributed network of Institutional Biosafety Committees. The NIH–RAC evolved a series of guidelines for recombinant DNA research that has become the standard for contained laboratory experimentation.

After a decade of successfully using guidelines and institutional oversight, the technology followed its normal sequence of activities leading to small–scale testing to be conducted outside of laboratory containment. Quite independently, but coincidentally, the National Research Council (NRC) published the "Redbook" (NAS, 1983) which set out a new paradigm for risk

---

*The views expressed are not necessarily those of the U.S. Department of Agriculture.

analysis. It is important to understand the dimensions of the NRC risk paradigm because it directly influenced subsequent policy decisions regarding the regulation of biotechnology.

PRINCIPLES OF RISK ANALYSIS

The risk paradigm provided in the "Redbook" described the process of risk analysis as being made up of:

Risk Assessment
Risk Management
Risk Communication

The sequential steps in risk assessment are: the identification of a hazard, followed by an assessment of exposure, and then risk characterization. Exposure is made up of fate and effects, when the focus of the assessment is on the environmental release of an organism. Conceptually taken together the identification of a hazard *times* the exposure is the characterization of a risk, or:

Risk Characterization = Hazard x Exposure,
when
Exposure = Fate x Effects

Risk assessment should be conducted with a sound scientific basis and use inferences as appropriate.

Risk management is the process of determining what to do about a characterized risk. This includes risk prevention, as well as the identification, selection and use of mitigating measures to reduce risk. Environmental risk management considerations often include social, economic and political judgements. The process of risk management should be institutionally separated from risk assessment.

Risk communication is an interactive process that promotes the exchange of information and opinions about risk among individuals, groups and institutions. This process should include providing access for stakeholders, or participation by and appreciation of public perceptions of risks.

THE FEDERAL COORDINATED FRAMEWORK

As the process of biotechnology research approached small–scale testing outside of contained laboratories, the Executive Branch of the Federal government began extensive discussions on how to coordinate regulatory activities to assure adequate protection of public health and the environment *vis–a–vis* biotechnology (ca. 1984). One of the foundations of the Federal Coordinated Framework for Regulation of Biotechnology (OSTP, 1986) is that the regulatory decision should be risk–based, and thus was set in motion the process of applying the "Redbook's" principles for risk analysis.

The first step in risk analysis is the identification of a hazard. It was clearly evident to the Federal Coordinated Framework policymakers that not all biotechnology represented a hazard. For instance, the application of somaclonal variation to crop improvement was definitely biotechnology, but it did not represent an unusual hazard. Additionally, some activities in recombinant DNA biotechnology were accepted as no, low or reasonable risk, and therefore these were not prime candidates for regulation.

What was identified was the fact that some *products* of biotechnology may represent an unusual hazard. Thus, these products should be the subject of risk assessment and regulation. It was therefore asserted that the products of biotechnology, not the process, should be the focus of Federal biotechnology regulation.

Risk assessment should be conducted with a sound scientific basis and use inferences as appropriate.

This distinction eventually became misconstrued by a few zealots of the Federal Coordinated Framework for Regulation of Biotechnology. They took the principle one step further, in an inept attempt to obliterate the distinction between conventional methods and the new biotechnologies. A terminology mind–game reminiscent of the popular book *1984* (Orwell, 1949), dominated Federal regulatory terminology during the Bush Administration. The use of common scientific terms like "genetically modified organisms," "transgenic" and "genetically engineered," was forbidden in regulatory language. As policy was derived from the Federal Coordinated Framework, some creative terminology had to be invented by technical editors (e.g.,"deliberately modified hereditary traits") to comply with these policies.

Another primary principle of the Federal Coordinated Framework was that there would be no new laws. This principle was derived from the assumption that existing statutory authorities were sufficient for Federal regulatory agencies to regulate the products of biotechnology. In the beginning this made a lot of sense. The first applications of biotechnology were emerging as pharmaceuticals and drugs, and this industry had long been regulated. Thus, there was considerable resistance on the part of the drug industry to biotechnology regulation *per se*, inasmuch as the existing regulatory structure seemed clearly sufficient.

This was not however the case for much of the rest of the applications of biotechnology, especially in agriculture. Much of the agriculture research enterprise had never been regulated, and existing authorities to deal with clearly identified special threats (such as plant pests; pesticides; toxic substances; animal viruses, serums and toxins; meat and poultry inspections; and food additives) stretched this eclectic collection of authorities over the domain of agricultural biotechnology. This became a challenge that has pretty much been met successfully. This success was achieved through the promulgation of new regulations under existing legislative authorities, individual efforts and a commendable level of interagency coordination to get the job done.

In spite of the extraordinary efforts to make the Federal Coordinated Framework fit the structure of agricultural biotechnology, some gaps and overlaps still exist. For example, it is still not clear how transgenic fish will be regulated, either in the research stage or as commercial products. U.S. Federal authority for the regulation of aquatic species is yet to be resolved, and as a consequence there is no existing statutory authority for fish and shellfish biotechnology products. Similar situations exist for non–pest insects, amphibians, reptiles, plants that have been transformed with sequences not from plant pests, and non–pest and non–pesticidal microorganisms when the research has no commercial intent (e.g., some types of university research with rhizobium). Moreover, the final promulgation of the Environmental Protection Agency (EPA) authorities under the Federal Insecticide, Fungicide, and Rodenticide Act (FIFRA) and the Toxic Substances Control Act (TSCA) has yet to take place.

If the regulation of biotechnology is to be risk-based, a clear and general agreement on what constitutes a hazard needs to be reached.

Given the above considerations some have called it remarkable that the Federal Coordinated Framework has been so successful for agricultural biotechnology regulation. A lot of the credit for this success goes to a few people in the Federal regulatory agencies that have given extraordinary effort to make it succeed.

THE NEXT GENERATION

As the technology progresses through its normal sequence, many of the products of agricultural biotechnology are ready for larger–scale performance testing, pre–commercial evaluations and eventual commercialization. As a consequence there are considerations that go beyond small–plot testing during these subsequent stages of product development that will place further strain on the processes of the Federal Coordinated Framework. Certainly the experiences and knowledge gained from small–scale testing can be used to better predict performance in larger–scale testing and commercial use, but not everything is directly translatable. The identification of the hazards of large–scale testing, the consideration of exposure, the numerics of large populations, and the probability values for fate will all take on new dimensions in large–scale tests. The question now being asked is: Is the Federal Coordinated Framework for Regulation of Biotechnology fully adequate to address today's and tomorrow's questions regarding the risks of biotechnology?

As the question on the Federal Coordinated Framework's adequacy is asked, is this the time to revisit the fundamental principles upon which it rests?

The regulatory issue of product versus process is not truly a settled issue, at least in the minds of many. The process of biotechnology and the transforming of organisms with foreign DNA represents to many an identified hazard requiring risk analysis. This perspective is no doubt related to the U.S. Food and Drug Administration's (FDA) May 1992 request for more public comment on its policy regarding the labeling of foods derived from new plant varieties. The FDA wants to know:

—Should all foods derived from new plant varieties developed using "genetic engineering" techniques be required to be labeled as such?
—Should labeling the source of introduced DNA be required?
—Under what circumstances is ingredient labeling appropriate?
—How can required labeling for food allergies be accomplished?
—What are the practical difficulties and economic impacts of labeling "genetically engineered" foods?

Clearly the regulatory issue of hazard identification for biotechnology is not resolved. Different perspectives on what constitutes a hazard complicates the development of a consensus. Without completing the first step of risk assessment, the application of scientific objectivity to the rest of the process will not be sufficient for those with opposing views.

WHAT IS NEEDED?

If the regulation of biotechnology is to be risk–based, a clear and general agreement on what constitutes a hazard needs to be reached. This involves reconciliation of the different views of the world where commercial interests advise the use of marketplace determinants; regulators prefer the use of a hierarchial, authoritative decisionmaking process; scientists assert the need for a rational process; and those concerned for the environment wish to apply a natural standard to the identification of a hazard.

Adding further complication is how different standards of objectivity are judged. Scientific objectivity is based on standardized techniques which permit experimental reproducibility. Social inquiry studies are considered objective if devoid of personal bias. Legal proceedings are considered objective if the participants adhere to the principle of disinterestedness. A lawyer would hardly view a scientist as objective if the scientist has an interest in that brand of science. Conversely, a scientist would accuse the lawyer of being subjective if the judgements were not truly reproducible. Social scientists share similar concerns for jurisprudence and biological science as they do not see them as necessarily free of personal bias. Who then is to provide objective judgements for biotechnology regulation?

Until we can resolve the issues of what constitutes a biotechnology hazard and who will make the objective judgements, it is not very likely that a consensus will emerge on how to proceed.

CONSENSUS BUILDING

Figure 1, taken from Douglas and Wildavsky (1983) presents the four problems of consensus building in risk analysis. In each of the four cells there is represented a risk problem, and a proposed solution. The dimensions of the block are knowledge and consent. In the upper left cell knowledge is certain and consent is complete. If there is a technical problem, it is merely a matter of making a calculation to derive a solution.

In other circumstances knowledge is uncertain, although consent is complete (upper right cell). In these situations the problem is not enough

FIGURE 1

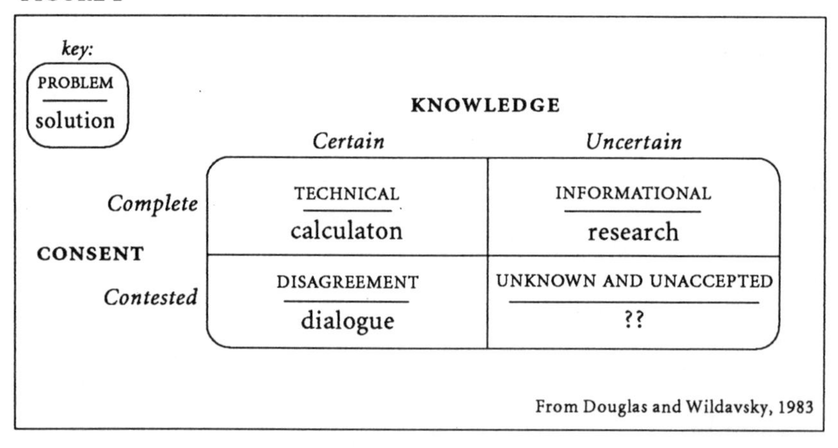

From Douglas and Wildavsky, 1983

information and the indicated solution is to conduct risk assessment research to resolve the problem. This is the approach that the U.S. Department of Agriculture (USDA) has implemented through Section 1668 of the 1990 Farm Bill. The USDA sets aside 1.0 percent of its biotechnology research outlays to conduct risk assessment research to fill knowledge gaps. This year the Department will award competitively $1.7 million for risk assessment research projects to help facilitate science–based, regulatory decisionmaking.

In those cases where knowledge is certain, but consent is contested, the problem becomes one of disagreement and the solution is dialogue (lower left cell). The role of organizations such as the National Agricultural Biotechnology Council (NABC) is very important in this type of circumstance. Through public dialogue, understanding can be built on existing, certain knowledge and perspectives of divergent views being more clearly understood.

It is the fourth cell (lower right cell) that represents the most difficult situation. In this case knowledge is uncertain *and* consent is contested. Presently, there is not a good solution for this type of situation. Clearly dialogue would be desirable. And more research might help resolve the unknown. But bringing together the information and divergent views for problem resolution represents one of the most tricky responsibilities facing science, government, public and private institutions, and the concerned public.

WHAT IS BEING DONE?

Discussions are now underway as to the adequacy of the Federal Coordinating Framework for Regulation of Biotechnology. As research discoveries progress along the path to commercialization, closing the information gaps, eliminating unnecessary duplication, and deregulating technologies known to be safe seem to be important items for our national agenda. Some of these changes can be seen with the recent implementation of an Animal and Plant Health Inspection Service (APHIS) notification and petition process for plant biotechnology regulation under the Plant Pest Act authority.

Also, the USDA's Marketing and Inspection Service is meeting with the FDA to coordinate new regulations for meat and poultry biotechnology and to close the gap for fish and shellfish. This latter gap may however require new legislation and this would represent a departure from a major principle embedded in the Federal Coordinated Framework. Other existing regulatory gaps may require similar gap–filling legislation.

There is clearly a need for more biosafety information exchange, both nationally and internationally.

There is clearly a need for more biosafety information exchange, both nationally and internationally. The Stockholm Environmental Institute recently established a free advisory service for Third World countries wishing an evaluation of the safety of field tests with genetically modified organisms. U.N. agencies are now looking at their role in information support systems for biotechnology on a global basis. The Organization of Economic Cooperation and Development is continuing to provide international leadership on the principles of biotechnology regulation and safety, and the European Economic Committee is working with the U.S. government to coordinate research and regulatory activities.

A major issue in these efforts is the extent to which the U.S. scientific community can and should become involved in the international exchange of information. If biotechnology is expected to be a major advantage in U.S. competitiveness in global markets, how much information should the U.S. share, given the advantage of a technological lead? Opportunities to collaborate in biotechnology risk assessment, biosafety data exchange, and regulatory practices seem to be obvious areas for international collaboration for mutual benefit. But to be successful, greater organizational and financial support for such international collaboration is needed.

There is a clear need for national educational programs in biotechnology that would organize factual information to be shared with the interested public, both youth and adult. The USDA's National Biological Impact Assessment Program is sponsoring a pilot project with the University of California,Davis, targeting school–age children with teaching materials on the scientific principles of biotechnology. The Agricultural Research Institute is looking for partners to assist in the development of a biotechnology educational program focused on adults. These are but two examples of what has been started in biotechnology public education.

Finally, there is a need for more national biotechnology dialogue, as there is not so much the absence of factual knowledge, but clearly different views on values, standards and preferences. NABC should continue to play an important role by providing a forum for continued biotechnology dialogue that will hopefully diminish disagreements and build toward a consensus on a national direction for agricultural biotechnology.

REFERENCES

Office of Science and Technology Policy (OSTP), Executive Office of the President. 1986. Coordinated Framework for Regulation of Biotechnology; Announcement of Policy on Notice for Public Comment. *Federal Register.* 51 (June 26): 23302–23393.

Douglas, M. and A. Wildavsky. 1983. *Risk and Culture.* University of California Press, Berkeley, CA.

National Academy of Sciences (NAS). Committee on the Institutional Means for Assessment of Risks to Public Health, Commission on Life Sciences. 1983. *Risk Assessment in the Federal Government: Managing the Process.* National Research Council. National Academy Press, Washington, DC.

Orwell, G. 1949. *1984.* New American Library, New York, NY.

Risk Assessment: A Technical Perspective

Roy L. Fuchs
Manager,
Regulatory Sciences,
Monsanto Corp.
with Terry B. Stone
and Paul B. Lavrik
Monsanto Corp.
(Fuchs pictured
on left)

REGULATORY JURISDICTION: GENETICALLY ENGINEERED PLANTS

After more than ten years of plant biotechnology research and extensive field evaluation at sites throughout the United States (Gasser and Fraley, 1989), the first of several improved crop products being developed are undergoing regulatory review prior to market introduction. Before these products were submitted for review, the various U.S. regulatory bodies had considerable input and oversight into the research and development undertaken to verify the performance and safety of these products. Initial research in the laboratory was performed under the National Institutes of Health (NIH) Recombinant DNA Guidelines (1976; 1982). Several years of field–testing were carried out under the jurisdiction of the United States Department of Agriculture (USDA), in conjunction with the U.S. Environmental Protection Agency (EPA) for pesticidal plants. Regulatory approvals may be required by all three regulatory agencies (USDA, EPA and FDA [Food and Drug Administration]) prior to marketing these products. Early in this regulatory process, the Office of Science and Technology Policy (OSTP) issued the "Coordinated Framework for Regulation of Biotechnology" (1986) which assigned appropriate regulatory oversight and responsibility for biotechnology to these federal agencies. (see discussion by MacKenzie, p. 47)

USDA regulates the movement and release of genetically engineered plants under the Plant Pest and Quarantine Act since many of these plants are generated using organisms or DNA sequences from organisms that are plant pests. The program is administered by the Animal and Plant Health Inspection Service (APHIS). APHIS will make a determination concerning the plant pest status of varieties derived via biotechnology prior to market introduction. EPA regulates the safety of the plant pesticidal products (e.g., proteins that provide insect or disease resistance) as active ingredients and the protein(s) used in the selection process for generating these plants (e.g., selectable marker proteins) as inert ingredients. FDA oversees the food and feed safety of products derived from these plants.

During the past decade, as progress has been made in the production, testing and development of plant biotechnology products, regulations and safety assessment requirements have developed to assure the safety of these products. There has been significant progress in the past twelve months by each of the three regulatory agencies in providing guidance and requesting public feedback on how these products will be regulated. The USDA published draft guidelines in June of 1992 (APHIS, 1992a) for public comment. The final policy was published in March, 1993 (APHIS, 1993). EPA published their proposed policy on regulating plant pesticides, under the Federal Insecticide, Fungicide and Rodenticide Act (FIFRA), for public comment and sponsored a Science Advisory Panel discussion on December 18, 1992 to obtain public input on this proposed policy (EPA, 1992). Likewise, FDA published their proposed policy for regulating genetically engineered plants and plant products on May 29, 1992 for public comment (FDA, 1992).

The food, feed and environmental safety of each of these will be assessed prior to marketing.

The first of the products produced by plant biotechnology, the Flavr Savr™ tomato by Calgene, Inc. (Sheedy et al., 1988), has been reviewed for plant pest status by the USDA (APHIS, 1992c) and is presently under review for food and feed safety by the FDA (Redenbaugh et al., 1992). A petition to the USDA has also been submitted for virus–resistant squash by the Upjohn Company (APHIS, 1992b). With over 500 field tests that have been performed around the world to date (Huttner et al., 1992; Casper and Landsmann, 1992) many other products are being extensively evaluated for agronomic, environmental and consumer value. The food, feed and environmental safety of each of these will be assessed prior to marketing. One of these, a potato improved to control a specific insect pest without the use of chemical pesticides, serves as a case study for how this safety assessment is being performed to assure product wholesomeness and safety, environmental soundness and to support public confidence in these products.

COLORADO POTATO BEETLE RESISTANT POTATOES

Potato plants that control that crop's most serious insect pest, the Colorado Potato Beetle (CPB), are among the first products that Monsanto has developed and for which we are completing a safety assessment. We have worked for nearly 10 years to develop and evaluate these potato plants, which resist CPB through the production of an insect control protein, found in nature, that selectively controls the beetle without affecting nontarget insects, humans or animals. These plants were produced by inserting and expressing a gene from *Bacillus thuringiensis* subsp. *tenebrionis* (*B.t.t.*) in the potato plant (Perlak et al., 1993) using *Agrobacterium* transformation (Newell et al., 1991). The CPB–resistant plants produce low levels of two new proteins, the *B.t.t.* protein for resistance to CPB and the neomycin phosphotransferase (NPTII) protein produced to enable the selection of cells expressing the *B.t.t.* protein in tissue culture. The *B.t.t.* protein produced in these potato plants is the same as one of the insecticidal proteins contained in several microbial formulations that have been used safely and have been commercially available since 1988 (EPA, 1988).

CPB [Colorado Potato Beetle] is the most damaging pest of the $2.3 billion U.S. potato crop...

CPB is the most damaging pest of the $2.3 billion U. S. potato crop (Casagrande, 1987; National Potato Council, 1992) and economically important in the majority of the North American potato production regions. Loss of revenue in Michigan alone was estimated at more than $15 million in a state where potato production is valued at $70 million (Potato Growers of Michigan, 1992; Olkowski et al., 1992). If untreated or poorly managed, the CPB can devastate potato production in some areas (Hare, 1980; Ferro et al., 1983; Shields and Wyman, 1984). Current treatment of CPB primarily involves the use of insecticides that are variably effective due to environmental factors and insect insensitivity, and significantly reduce field populations of beneficial insects which help control other potato pests. These pesticides are also expensive, with costs that can exceed $200 per acre per season (Ferro and Boiteau, 1992).

Field trials conducted in 1991 with CPB–resistant potatoes demonstrated effective control of feeding damage by all stages of the CPB. There were significantly fewer immature larvae, adults and egg masses of CPB on the genetically improved potatoes, compared to the control plants. Without insecticide application, defoliation of the improved potato plants was less than, or equal to, control plants sprayed with insecticides on a regular schedule. In addition, agronomic evaluations consisting of plant vigor, growth habit characteristics and general insect and disease susceptibility, have shown the CPB–resistant potatoes to be equivalent to the parental Russet Burbank potatoes. Field tests were expanded in 1992 with similar results.

These genetically improved potatoes offer several advantages to the grower, the consumer and the environment for controlling this devastating insect pest. The superior CPB control offered by the plants will enable growers to significantly reduce the amount of chemical insecticide now applied to their crop while maintaining comparable yields. Reducing the amount of insecticide applied to potatoes will further aid the implementation of Integrated Pest Management (IPM) practices as beneficial insect populations will be maintained, which can help reduce the other pests of potatoes not directly controlled by the CPB–control protein. The *B.t.t.* protein also has been shown to be safe to nontarget species, including humans (EPA, 1988) and thus provides an environmentally safe means to control CPB. In addition, CPB–resistant potatoes will benefit both large and small growers as no additional labor, planning or machinery is required for adoption. Prior to market introduction, the potato lines will continue to be evaluated for performance and to refine insect management programs. Efforts will focus on confirming CPB control across the potato growing regions and developing production systems that optimize the benefits of these improved potatoes.

The first approach is to provide appropriate information to establish that the CPB-resistant potatoes are "substantially equivalent" to the Russett Burbank potatoes...

As with other food crops improved through biotechnology, USDA, EPA and FDA exercise joint regulatory oversight for these genetically improved potatoes. All field tests have been carried with the approval of the USDA. In October 1992, an Experimental Use Permit (EUP) was requested from the EPA. The EUP permit, approved in May, 1993, allows for more extensive field evaluation of CPB–resistant potato varieties to be performed on more than 10 acres. Prior to commercialization, a USDA determination of the nonregulated status and an EPA product registration will be obtained under the respective policies described above. Likewise, appropriate consultations with and oversight by FDA will be conducted as described in the FDA policy.

SAFETY ASSESSMENT OF CPB–RESISTANT POTATOES

Monsanto's policy for ensuring the safety of the CPB–resistant potatoes is consistent with the published policies of the three regulatory agencies and relies on two approaches. The first approach is to provide appropriate information to establish that the CPB–resistant potatoes are "substantially equivalent" to the Russet Burbank potatoes from which this variety was derived. The term "substantial equivalence" refers to the concept that the genetically improved potatoes are comparable to the Russet Burbank potatoes in respect to composition, nutritional quality, yield, morphology and in other aspects that could impact the use, value, and the environmental, food and feed safety of this product. The only significant difference that has been observed between the genetically improved potatoes and the Russet Burbank potatoes is that the CPB–resistant potatoes effectively control the insect pest by expressing two additional proteins (*B.t.t.* and NPTII). No other differences have been observed. The concept of "substantial equivalence" or "substantial similarity" has been used by FDA (1992) and other international organizations (Organisation for Economic Co–operation and Development, 1992; International Food Biotechnology Council, 1990) in their recommended approaches for safety assessment. In essence, this is the way new plant varieties and plant products have traditionally been regulated. In addition to establishing the "substantial equivalence" of the CPB–resistant potatoes to the Russet Burbank potatoes, we will also provide data to confirm the environmental, human and animal safety of the two newly expressed *B.t.t.* and NPTII proteins.

SUBSTANTIAL EQUIVALENCE OF CPB–RESISTANT POTATOES TO RUSSET BURBANK POTATOES

Information concerning the source(s) of the genes introduced, the methods used to produce the genetically improved potato plants, the molecular characterization of DNA introduced into these plants, and characterization of the levels of the *B.t.t.* and NPTII proteins serves as a basis for characterizing the CPB–resistant plants. The important nutritional and natural products in potato are being determined for both the genetically improved and Russet Burbank potatoes to show that the composition of the potato tuber has not been altered during the transformation and regeneration processes. Levels of the macronutrients—protein, fat, carbohydrate, dietary fiber and ash—are being determined. The levels of the important vitamins—vitamin C, vitamin B_6, thiamine, niacin, folic acid and riboflavin—are being assessed. Levels of important minerals—calcium, copper, iron, iodine, magnesium, phosphorus, potassium, sodium and zinc—are being evaluated. The only class of important potato natural toxicants, the glycoalkaloids (solanine and chaconine), have been quantified and shown to be comparable in both the genetically improved and Russet Burbank potatoes. Raw potato tubers from both the CPB–resistant and Russet Burbank potatoes were fed, along with the regular rat diet, to rodents in a 28–day study to assess the palatability of these potatoes. No differences in consumption, growth rates, or observations during gross necropsy were observed during these studies. These data have confirmed that the CPB–resistant potatoes are comparable to the Russet Burbank potatoes in all aspects except for the ability to control the CPB pest due to the presence of minor amounts of *B.t.t.* and NPTII proteins.

...use of registered* B.t.t. *products offers no significant risks to human health or nontarget organisms.

SAFETY OF THE *B.t.t.* PROTEIN

The *B.t.t.* gene used to produce CPB–resistant potato plants, designated *cryIIIa*, (Hofte and Whiteley, 1989) was isolated from DNA from *Bacillus thuringienses* subsp. *tenebrionis* (McPherson et al., 1988). The *B.t.t.* gene encodes an insecticidal protein produced by these bacteria during sporulation. The protein is selectively active against coleopteran larvae. Upon ingestion by susceptible species, feeding is inhibited with disruption of the gut epithelium and eventual death of the insect pest (Slaney et al., 1992). The amino acid sequence encoded by the gene inserted into potato plants produces a protein that is identical to that produced by *B.t.t.* found in nature (McPherson et al., 1988).

Based on the available scientific data, EPA and other regulatory agencies worldwide have determined that use of registered *B.t.t.* products offers no significant risks to human health or nontarget organisms (Shields and Wyman, 1984; EPA, 1991). In published reviews and the EPA documents, studies are referenced where large doses (5000 mg per kg) of *B.t.t.* preparations were administered as single or multiple doses to different laboratory animals with no adverse effects. Avian and aquatic organisms have also been fed *B.t.t.* preparations with no adverse effects. The preparations which were administered contained varying amounts of crystalline proteins from *B.t.t.*, either as a mixture with spores or encapsulated in killed *Pseudomonas fluorescens* cells (EPA, 1991). While target insects are susceptible to oral doses of *B.t.t.* proteins (μg per gram of body weight), there was no evidence of any toxic effects observed in nontarget laboratory mammals, including fish or birds given the equivalent of up to 10^6 μg of protein per gram of body weight. No deleterious effects were observed on nontarget insects at doses over 300– to 700–fold that needed to control the target insects (MacIntosh et al., 1990). In addition to the predicted lack of receptors for the *B.t.t.* protein, the absence of adverse effects in nontarget animals is further facilitated by the poor solubility and rapid degradation of *B.t.t.* proteins in the acid environment of the digestive system.

To confirm the safety of the *B.t.t.* protein expressed in CPB–resistant potatoes, we have obtained gram quantities of purified *B.t.t.* protein by expressing this protein in microbial systems (*E. coli*). Limited expression of this protein prohibited the isolation of large quantities of this protein from the potato tubers or potato plant directly. Minor amounts of this protein, purified from the potato tuber and from microbes, were shown to be chemically and functionally equivalent. A series of commonly used analytical assays were used for this equivalence assessment. An acute gavage study was conducted in mice to confirm the safety of the *B.t.t.* protein. A dose, following EPA guidelines, was used that was equivalent to over 2.5 million–fold safety factor based on the average consumption of potato and the level of the *B.t.t.*

protein present in the tuber. No adverse effects were observed in terms of food consumption, weight gain, mortality or gross necropsy observations. Purified protein was also used in an *in vitro* digestion experiment which demonstrated that the *B.t.t.* protein has an extremely short half–life (less than 20 seconds) under simulated gastric conditions (The United States Pharmacopeia, 1990). These studies confirm the safety of the *B.t.t.* protein to humans and animals.

The specificity of the *B.t.t.* protein to CPB was confirmed by host–range studies using the purified *B.t.t.* protein. Five nontarget, beneficial insects (including honey bees, lacewings, ladybird beetles and a parasitic wasp) were also shown to be unaffected by doses of the purified *B.t.t.* protein that are greater than 100 times the amount required to affect the CPB–target insect. These studies confirm the specificity of the *B.t.t.* protein and the safety of this protein to nontarget insects. We are also performing studies to confirm the rapid degradation of the *B.t.t.* protein in the soil after potato tubers are harvested.

SAFETY OF THE NPTII PROTEIN

The NPTII protein, and hence the gene encoding this protein, was used as a selectable marker to enable the identification of potato cells that contained the *B.t.t.* gene. The description and safety assessment of the NPTII protein has been discussed in detail in the FDA submission for an advisory opinion by Calgene (1990) and by recent articles by Flavell et al. (1992) and Nap et al. (1992). In addition to this information, we have performed similar equivalence acute gavage and digestive fate studies as described for the *B.t.t.* protein above, with similar results. No adverse affects were observed in the acute gavage study with greater than a 5 million–fold safety compared to projected consumption, and the half–life of the NPTII protein in the simulated digestive fate study was also less than 20 seconds, confirming the mammalian safety of this protein.

SUMMARY

A variety of plant biotechnology products have been developed and extensively tested under field conditions. Appropriate regulatory oversight has evolved, and continues to evolve, through the various stages of the development of this technology. All three regulatory agencies in the U.S. issued either draft or final policies during 1992 that outline their policies on regulating genetically engineered plants. Two plant products are currently being reviewed under these policies and several more are expected to follow closely behind. One of these is the CPB–resistant potatoes that is described in this report. We have described the approaches that we are using to assess the food, feed and environmental safety of this product as an example. The philosophy is based on the concept of "substantial equivalence." The new potato variety, derived using plant biotechnology, is established to be comparable to traditionally bred potatoes by comparing nutritional quality, level of important natural products, and agronomic and environmental performance. In addition, a direct safety assessment of the newly expressed proteins (*B.t.t.* and NPTII, for the CPB–resistant potato) confirmed the safety of these components. These safety assessments have confirmed the food, feed and environmental safety of the CPB–resistant potatoes. Similar assessments are being performed for other plant varieties derived using plant biotechnology to assess the safety of these products.

ACKNOWLEDGMENTS

We thank Steve Rogers and Frank Serdy for critical review of this report and thank the member of the Insect Resistant Potato Technical and Regulatory Science teams who have and are in the process of developing the safety data package on this product for their excitement, support and outstanding technical accomplishments.

REFERENCES

Animal and Plant Health Inspection Service (APHIS). 1993. Genetically Engineered Organisms and Products; Notification Procedures for the Introduction of Certain Regulated Articles; and Petition for Nonregulated Status; Final Rule. *Federal Register.* 58:17044.

Animal and Plant Health Inspection Service (APHIS). 1992a. Genetically Engineered Organisms and Products; Notification Procedures for the Introduction of Certain Regulated Articles; and Petition for Nonregulated Status. *Federal Register.* 57:53036.

Animal and Plant Health Inspection Service (APHIS). 1992b. Notice of proposed interpretive ruling in connection with the Upjohn Company petition for determination of regulatory status of ZW020 virus resistant squash. *Federal Register* 57:40632.

Animal and Plant Health Inspection Service (APHIS). 1992c. Proposed Interpretive Ruling in Connection with Calgene, Inc. Petition for Determination of Regulatory Status of Flavr Savr™ Tomato. *Federal Register.* 57:31170.

Calgene, Inc. 1990. *Request for Advisory Opinion—kan^r Gene: Safety and Use in the Production of Genetically Engineered Plants.* FDA docker Number: 90A–0416.

Casagrande, R.A. 1987. The Colorado Potato Beetle: 125 Years of Management. *Bull. Entomol.Soc.* 33:142.

Casper, R. and J. Landsmann, eds. 1992. The 2nd International Symposium on: *The Biosafety Results of Field Tests of Genetically Modified Plants and*

Organisms. Biologische Bundesanstalt fur Land– und Forstwirtschaft, Braunschweig, Germany.
Environmental Protection Agency (EPA). 1992. EPA Proposal to Clarify the Regulatory Status of Plant–Pesticides. *Federal Register*. 57:55531.
Environmental Protection Agency (EPA). 1991. Delta Endotoxin of *Bacillus thuringiensis* variety *san diego* Encapsulated in Killed *Ps fluorescens*. *EPA Pesticide Fact Sheet*, EPA/OPP Chemical Code Number 128946–1.
Environmental Protection Agency (EPA). 1988. Guidance for the Reregistration of Pesticide Products Containing *Bacillus thuringiensis* as the Active Ingredient. U.S. Goverment Printing Office. NTIS PB 89–164198.
Ferro, D.N. and G. Boiteau. 1992. Management of Major Insect Pests of Potato. In *Plant Health Management in Potato Production*. R.C. Rowe, ed. American Phytopath. Soc. Press, St. Paul, MN. p. 103–115.
Ferro, D.N., B.J. Morzuch and D. Margolies. 1983. Crop Loss Assessment of the Colorado Potato Beetle (*Coleoptera: Chrysomelidae*) on Potatoes in Western Massachusetts. *J. Econ. Entomol.* 76:349.
Flavell, R.B., E. Dart, R.L. Fuchs and R.T. Fraley. 1992. Selectable Marker Genes: Safe for Plants? *Bio/Technology*. 10:141.
Food and Drug Administration (FDA), Department of Health and Human Services. 1992. Statement of Policy: Foods Derived from New Plant Varieties. *Federal Register*. 57:22984.
Gasser, C.S. and R.T. Fraley. 1989. Genetically Engineering Plants for Crop Improvement. *Science*. 244:1293.
Hare, J.D. 1980. Impact of Defoliation by Colorado Potato Beetle on Potato Yields. *J. Econ. Entomol.* 73:369.
Hofte, H. and H.R. Whiteley. 1989. Insecticidal Crystal Proteins of *Bacillus thuringiensis*. *Microbiol. Rev.* 53:242.
Huttner, S.L., C. Arntzen, R. Beachy, G. Breuning, L. De Francesco, E. Nester, C. Qualset and A. Vidaver. 1992. Revising Oversight of Genetically Modified Plants. *Bio/Technology*. 10:967.
International Food Biotechnology Council (IFBC). 1990. Biotechnologies and Food: Assuring the Safety of Foods Produced by Genetic Modification. *Register Toxicology Phamacol.* 12:S1.
MacIntosh, S.C., T.B. Stone, S.R. Sims, P.L. Hunst, J.T. Greenplate, P.G. Marrone, F.J. Perlak, D. A. Fischhoff and R.L. Fuchs. 1990. Specificity and Efficacy of Purified *Bacillus thuringiensis* Proteins against Agronomically Important Insects. *J. Invert. Path.* 56:258.
McPherson, S.A., F.J. Perlak, R.L. Fuchs, P.G. Marrone, P.B. Lavrik and D.A. Fischhoff. 1988. Characterization of the Coleopteran–Specific Protein of *Bacillus thuringiensis* subsp. *tenebrionis*. *Bio/Technology*. 6:61.
Nap, J.P., J. Bijvoet and W.J. Stikema. 1992. Biosafety of Kanamycin–Resistant Transgenic Plants: An Overview. *Transgenic Crops*. 1:239.

National Institutes of Health (NIH). 1982. Guidelines for Research Involving Recombinant DNA Molecules. *Federal Register.* 47:38048.

National Institutes of Health (NIH). 1976. Recombinant DNA Research Guidelines. *Federal Register.* 41:27902.

National Potato Council. 1992. *Potato Statistical Yearbook.* Englewood, CO.

Newell, C., R. Rozman, M. Hinchee, E. Lawson, L. Haley, P. Sanders, W. Kaniewski, N. Tumer, R. Horsch and R. Fraley. 1991. *Agrobacterium*–Mediated Transformation of *Solanum tuberosum* L. cv. 'Russett Burbank'. *Plant Cell Reports.* 10:30.

Office of Science and Technology Policy (OSTP), Executive Office of the President. 1986. Coordinated Framework for Regulation of Biotechnology. *Federal Register.* 51(June 26):23302–23393.

Olkowski, W., N. Saiki and S. Daar. 1992. IPM Options for Colorado Potato Beetle. *The IPM Practitioner.* 16:1.

Organisation for Economic Co–operation and Development. 1992. *Safety Evaluation of Foods Derived by Modern Biotechnology: Concepts and Principles.* Paris.

Perlak, F., T.B. Stone, Y.M. Muskopf, L.J. Petersen, G.B. Parker, S.A. McPherson, J. Wyman, S. Love, D. Beaver, G. Reed and D. Fischhoff. 1993. Genetically Improved Potatoes—Protection from Damage by Colorado Potato Beetles. *Plant Mol. Biol.* 22:313–321.

Potato Growers of Michigan, Inc. and the Michigan Potato Industry Commission. 1992. *December 1991 CPB Survey Results: Crop Years 1989–1991.*

Redenbaugh, K., W. Hiatt, B. Martineau, M. Kramer, R. Sheehy, R. Sanders, C. Houck and D. Emlay. 1992. *Safety Assessment of Genetically Engineered Fruits and Vegetables: A Case Study of the Flavr Savr™ Tomato.* CRC Press, Boca Raton, FL.

Sheedy, R., M. Kramer and W. Hiatt. 1988. Reduction of polygalacturonase activity in tomato fruit by antisense RNA. *Proc. Natl. Acad. Sci. USA.* 85:8805.

Shields, E.J. and J.A. Wyman. 1984. Effect of Defoliation at Specific Growth Stages on Potato Yields. *J. Econ. Entomol.* 77:1194.

Slaney, A.C., H.L. Robbins and L. English. 1992. Mode of Action of *Bacillus thuringiensis* Toxin CryIIIA: An Analysis of Toxicity in *Leptinotarsa decemlineata* (Say) and *Diabrotica undecimpunctata Howardi* Barber. *Insect Biochem. Mol. Biol.* 22:9.

The United States Pharmacopeia. 1990. The United States pharmacopeial Convention, Inc., Rockville, MD. p. 1788.

Risk Assessment: A Farmer's Perspective

Will Erwin
Indiana Farmer
(center)

It is important to understand that this is only one farmer's perspective. I simply cannot speak for other farmers—who range from small, part–time farmers to large corporate farmers who have large professional staffs and many employees. I speak as an individual farmer reflecting the thoughts of a Midwest commercial family farmer who has had a number of responsibilities in state and federal government.

This presentation will discuss what farmers are like—really like—the comprehensive environment in which they operate, the macro changes in farmer decisionmaking, how farmers look at change in general, how farmers look at changes in biotechnology in particular, how farmers assess a new product and some of the issues we, as farmers, will be facing in the future.

It will highlight those points which tend to be overlooked about farmers and have particular relevance.

WHAT ARE FARMERS REALLY LIKE?

Farmers are well–educated people, averaging slightly more years of education than non–farmers, often with university degrees and frequently with master degrees and PhDs. As a group they work for less per hour than non–farmers, consume less and accumulate more than others. In short, they live poorer but die richer, but they do it because they want to for noneconomic reasons.

Among those noneconomic reasons are personal independence, love of and attachment to the soil, love of animals and nature, and a deep sense of stewardship. Most farmers put a high premium on religion. Daily working with the life and death realities of nature and isolation to think without interruption, increases religious commitment which the community discipline of rural people reinforces.

Farmers are increasingly anxious economically as they have felt the agricultural depression. They are increasingly uncomfortable about seemingly endless environmental hazards, be they perceived or real. Radon, the ozone layer and the unknowns of pesticides and biotechnology cause concern. Recent news stories of women with breast cancer having higher levels of DDT in their systems intensify concerns, both in the specific and in the future, about what their new information may indicate about all pesticides.

Perhaps farmers' most rapidly escalating hunger is for fact and truth, and they are less sure where to get it.

There is increasing fear of unreasonable regulation and even of entrapment—where farmers may follow all the rules and be found negligent, or where they may make the extra effort to be environmentally responsible and be found liable.

Farmers tend to trust their neighbors, their clergy, their farm organization, their university and extension people, as well as the business people they deal with. However, they are less comfortable with their government and the extremists who may influence government.

Increasingly, farmers are uncomfortable with agricultural leaders who take extreme antienvironmental positions, but they are also very concerned with unrealistic positions taken by some animal rightists and environmental spokespeople. Perhaps farmers' most rapidly escalating hunger is for fact and truth, and they are less sure where to get it.

WHAT IS THE ENVIRONMENT IN WHICH FARMERS OPERATE?

The knowledge explosion has left farmers increasingly awed by the realization that what they do know is a constantly reducing percent of the knowledge available. They feel a need for more knowledge and yearn for sources they believe are sound.

Farmers are increasingly vulnerable. A county judge once told me he could put anyone in the county in jail. There are so many laws, everyone is technically violating something, no matter how conscientious he or she is. This is compounded for the individual entrepreneurs who do not have professional staffs.

Farmers are misunderstood. The first real shock I had at the Environmental Protection Agency (EPA) was the reality that many fine, conscientious government employees were writing regulations for farmers while they themselves did not understand agriculture. For example, early on I was told by a fine, conscientious public servant who was writing regulations for farmers, that most of the farmland in the U.S. was owned by large corporations. (Farmers know that more than 90 percent of farmland is owned by families or individuals.) It is unreasonable for 98 percent of the population to be preoccupied with understanding the roughly 2 percent who farm. But the 2 percent who farm are the custodians of much of the surface of the earth, and unless reality is understood, everyone will lose.

The increasing sophistication of agricultural production technology in which biotechnology looms large raises increased questions as to how and if individual farms can function effectively without vertical integration or new systems to insure that the new technology is operative on smaller and medium–sized farms.

MACRO CHANGES IN DECISIONMAKING

Before discussing the changes in American farmers' decisionmaking, I want to point out that one of our great resources is that American farmers can make decisions. In my work in Bulgaria, I find that one of the major impediments to progress is that where people have had the State make business decisions for them for fifty years, the people have great difficulty in making the decisions required for doing business.

Based on my almost half century of farming, I would suggest the following as major changes in decisionmaking during the 20th century:

...many fine, conscientious government employees were writing regulations for farmers while they themselves did not understand agriculture.

The decisionmaking process is more complex due to increased information—some of which has to be inaccurate—increased and sometimes inconsistent regulations, and a decisionmaking climate of potential, and sometimes real, media–hyped anxiety.

Dependence on crop consultants, marketing consultants, management consultants, environmental consultants, feed consultants, accountants, lawyers and others to sort out the information avalanche has increased.

Farmers are less confident in decisions they make. Increased insurance—liability, pollution, health, and workman's compensation—reflect this. There is also some increase in the "I'll do my best and let the chips fall where they may" attitude.

There is more anxiety in the whole process of farming. Last week a county agricultural extension agent told me of a recent meeting on biotechnology in his area; he said people are really afraid of it. It appears to me that this fear typifies most current decisionmaking because:

—Scientific data are too complex for nontrained people to understand it;

—There is deep and vocal disagreement about the risk;
—Our culture hypes anxiety about the unknown;
—Farmers have been alarmed by past traumas such as DES, EDB and Alar;
—The rate in which science is disproving previous positions causes insecurity; and
—There is a substantial sense of regulatory harassment among farmers and anything new and complex bodes of more harassment.

HOW FARMERS LOOK AT CHANGE

Historically farmers have looked at change as exciting. This nation was settled by risk–takers who looked at the frontier as an opportunity to change their lives for the better while they made the wilderness more productive.

Currently, there is still the same excitement for change. Farm shows, demonstrations, field days and farm tours excite farmers as they see new things and concepts. But change is viewed with increased anxiety, feelings of vulnerability and sometimes even futility. Perhaps the shift is reflective of a general perception that rural discipline is shifting from a discipline based fundamentally on individual and community conscience to a discipline of government enforcement.

HOW FARMERS LOOK AT CHANGES IN BIOTECHNOLOGY

The initial response to how farmers look at changes in biotechnology is a combination of excitement and fear—*excitement* about the production potential; the hope of such things as genetic immunity reducing the losses from diseases and pests without the use of vaccines and pesticides, and *fear* that undesirable or even dangerous dimensions may be introduced. Farmers remember that the introduction of rabbits to Australia was supposed to be highly beneficial, and many of us here in Indiana had a hassle with multiflora rose which was to be a beneficial fence. But biotechnology carries a much higher fear level. Terms like "insecticidal protein" in corn create some anxiety as we are just now hearing more about the dangers of the pesticides used many years ago.

...the producer is in the middle faced with the reality that he has to decide while others debate.

There is further fear that genetic alterations may introduce risk to those with rare but intense allergies. For example, someone with a peanut allergy might now react to cornflakes made from transgenic corn containing a peanut protein.

There is further fear that something created by biotechnology might not be contained once released. DDT, EDB and Alar could be removed from the system, but a science fiction–type biological plague could escape and be "uncontrollable." I do not think this doomsday fear is very strong among farmers, but the 100 percent safe Delaney Amendment–type thinking has some appeal to everyone. There is some feeling that the traditional "nothing risked, nothing gained" philosophy should be rendered obsolete by science.

The initial response to how farmers look at changes in biotechnology is a combination of excitement and fear...

Following the initial response we find economic opportunity and anxiety. The hope of farmers to produce a larger and better product at a lower cost is universal, but the unknowns create anxiety. Some of these are:

—Will it create huge surpluses and break markets?

—If the U.S. regulates biotechnology, will the rest of the world run with it and take our foreign, and even domestic, markets?

—Will the big corporations monopolize the new products?

—Will it force vertical integration of farms? and

—Will it frighten consumers and destroy demand?

There is also what might be called the political–social fear. This is simply the discomfort of being caught in a whipsaw between differing societal and political action groups where no one is quite sure whom to believe, and the producer is in the middle faced with the reality that he has to decide while others debate.

HOW DO FARMERS ASSESS A NEW PRODUCT?

While farmers differ in systems and priorities in decisionmaking, most include the following questions:

—Is the new product safe? (Farmers have concerns about immediate toxicity, long–term health risk, immediate and long–term environmental risk and how reliable the safety measures are for its use.)

—Will it increase profitability if I use it, and will I be left behind if I do not?

—Will this product affect demand for what I produce positively or negatively?

—Does it fit in the systems of my farm? and

—Is it moral?

It is quite common to hear farmers say, "I don't want to use that stuff because it is too 'hot,'" or they do not want to use any chemicals they do not have to use because of residues and unknowns. In general, I think these same concerns are even greater regarding biotechnology.

WHAT ARE SOME OF THE ISSUES WE WILL BE FACING IN THE FUTURE? How will we get leaders to take the risk of leading? When I was still at EPA, I had a call from the president of a state farm group who said he was in big trouble because he had urged his farmers to be environmentally responsible and turn in their used oil for recycling rather than use it on the farm in a way that it might damage the environment. He said that about half followed his lead and they were now being held liable because the recycling plant had gone under and was a Superfund site, while the other half who had ignored him were home free. Policy officials at EPA were sympathetic, but the enforcement people were adamant, taking the attitude that "the law is the law." What is of particular concern was the number of knowledgeable people who, upon hearing of this problem, indicated that they were not surprised, and that it never pays to get out in front.

Risk is a price of progress. It must be assessed and managed.

How do we develop a realistic attitude toward risk? Risk, risk assessment, risk management and risk–to–benefit relationships have all consumed much of our thoughts. But logic does not grab human attention as much as fear does. The body politic wants simple brief explanations. Unfortunately, risk assessment at the citizens level is too often typified by the young mother who came to my wife during the Alar scare smoking a cigarette with her child in her arms and said, "Will apples hurt my baby?"

Progress and quality of life will be enhanced by our ability to focus on reality in relationship to risk and to communicate this reality to people in simple terms. Risk is a price of progress. It must be assessed and managed. Unperceived risks can do great damage, but non–risks perceived as risks retard progress. Whom the public will trust and how to communicate complex science to laypeople in simple terms are ongoing issues of increased urgency.

...one of our great resources is that American farmers can make decisions.

How to communicate realism about risk is particularly difficult in our democracy. The free enterprise system encourages competition, therefore, our people are bombarded with a "fear–fix" syndrome. TV commercials create insecurity about everything from bad breath to being cheated, so that they can sell security. News commentators and headline writers, competing for viewers and readers, each try to make their story the most exciting. Exaggerating risk is more exciting than cool analysis, and the limits of ethics are pressed. Politicians get elected by identifying risks they can fix, and they get little media coverage if they understate the risk. Some environmental extremists get prominence and contributions from extreme positions, and some agricultural extremists get prominence and contributions from extreme positions. Hard science and truth are often too complex, and, perhaps to the layman, too dull to attract much public interest until the issues are too polarized for easily reasoned solutions. All of this increases *fear*, and most have a *fix* to sell that is not as convincing as the fear. All of this makes the NABC meetings very important.

Finally, it is clear that farmers are uncomfortable about how much they need to know and with the recognition that they can never know enough. And they, like all the others, are evaluating whom to trust. We have lived through what I hope is the extreme of the antihero era, but not without damage to our most revered institutions. Unfortunately, some scar tissue remains, but credibility acceptance levels will recover slowly.

I am a product of the land–grant system and have profound and continuing respect for it. There is, however, a real need for our educational and research institutions to not only continue to look at their daily tactical need to survive during difficult times, but to examine in depth their strategic positions and set their sights on the horizon.

Many farmers have, over the years, received much of their thought stimulation both from their churches and the state university system. Some historically appreciated the theology of the church, but were somewhat "turned off" by the fundamentalist preoccupation with the evils of smoking, drinking and sexual promiscuity, while they were more inspired by the open–minded scientific approach of the university people.

...a real need for our educational and research institutions...to examine in depth their strategic positions and set their sights on the horizon.

In May, on the plane to Bulgaria, I read in the airline magazine a pragmatic article on communicative diseases which stated that the best cure for AIDS is to control sexual promiscuity. I then saw on CNN that the Senate was considering requiring warning labels on all alcoholic beverages. When this is added to the overwhelming evidence on smoking, I realized that those fundamentalists had been the most accurate in their positions, even though their views were arrived at through a theological rather than scientific analysis. When this is compounded by the concern farmers have when they read the current labels on pesticides and realize that the guidance given them in the past (which was the best science had to offer at the time) put them at risk by today's standards, there is real soul–searching.

In a cultural situation where individuals are increasingly overwhelmed by an explosion of information and made anxious in a culture that hypes fear, their increased anxiety and frustration may lead them to look to other than hard science for guidance.

This may seem unlikely, but when I was in India, I was amazed to see educated Indians defending the tradition of sending cows to old cows' homes, their carcasses to remain uneaten in a society abounding with protein–deficient children.

Frustrated and insecure people often reach out in unexpected ways. This is one more reason why this meeting which encourages an open dialogue on risk is so important.

Public Perceptions of the Benefits and Risks of Biotechnology

Thomas J. Hoban
Sociology and Anthropology, North Carolina State University
(pictured on left)
with Patricia Kendall
Food Science and Human Nutrition, Colorado State University

During the next decade, biotechnology is expected to have major impacts on food production and processing. Supporters predict significant economic, social and environmental benefits. Opponents raise serious concerns about risks and ethics of biotechnology. Biotechnology has become an important and controversial public policy issue that is drawing the attention of the media and the public. Social science research can provide industry, government, universities and others with valuable insights into public perceptions of biotechnology and related public policies. This paper presents an overview of the role of social science research and examines selected results of a recently completed project.

Most experts recognize that public knowledge and perceptions of biotechnology must be systematically evaluated. Stenholm and Waggoner (1992) explain how consumers will be the ultimate judge of emerging technologies. They will appraise the merits of a particular product and determine its success or failure. The Office of Technology Assessment (1992) points out that while many new technologies will soon be commercially viable, they will not automatically be put to use. The public increasingly questions whether technological change is good or needed. People are voicing new concerns about food safety, the environment and the changing structure of agriculture. Lack of public acceptance could prevent some technologies from being used even if they are approved by regulatory agencies.

Social science research can help design effective educational programs and public policies. Political, industrial and educational leaders need more information about potential public reaction to biotechnology. Consumers will make the ultimate decisions about the acceptability of food products developed through biotechnology through their market behavior. However, it is also useful to anticipate consumer reaction to the products of biotechnology before there has been a significant investment in research and development. Cross (1992) argues that government, academia and private industry must not wait until the questions are asked before information is provided to the public. It is necessary to identify different audiences and know how to reach them. Industry, government and universities must understand and respond to public opinion. Social science research helps make that possible.

Public participation will promote more effective and acceptable biotechnology policies.

Social science research also serves to broaden the debate to include more diverse perspectives. Much of the controversy surrounding biotechnology is not simply a matter of scientific facts and expert opinion. Some of the key issues revolve around the public's confidence in and ability to influence public and private decisions about the future of biotechnology. The Office of Technology Assessment (1992) notes that biotechnology is not so different from previous agricultural technologies as to raise novel scientific issues concerning the safety of foods. What is substantially different, however, is the climate in which this new class of technologies is being introduced. Society is increasingly skeptical of how new technologies are developed and regulated.

Social science research can serve as a valid and reliable mechanism for public participation. Public participation will promote more effective and acceptable biotechnology policies. Stiles (1989) argues that without adequate participation in decisions about biotechnology, the public will react as it has to other technologies. He explains how a technology must fit into society to maximize benefits and minimize social and political disruption. Public participation in society's decisions cannot be avoided. It either occurs in a planned and orderly fashion or in a reactive and disruptive fashion.

Social science research can also provide guidance to improve the design and implementation of educational programs. Public attitudes and knowledge must be researched, understood and considered before developing educational programs and communication efforts. Such research must illuminate the diverse types of information that are important to the public. Even after the issues have been identified, communication will not be easy or effective without systematic evaluation. Foreman (1990) explains the problems with assuming that communication will resolve public concerns about biotechnology. Differences over this issue may represent not a failure to communicate, but a conflict in values. Conflict also occurs if the risks and benefits of biotechnology do not accrue to the same individuals or groups. Social science research can help identify and evaluate the effectiveness of conflict management efforts.

Social science research helps better define the social and political context in which biotechnology is developing. The use of biotechnology in agriculture and food production could elicit food safety and environmental concerns similar to those expressed about agricultural chemicals. Other dimensions of biotechnology also draw public attention (e.g., socioeconomic impacts and ethical concerns). Senator Al Gore (1991) explains that what is needed to balance our technological prowess is a renewed engagement in the debate over biotechnology policy—not just the ethics of genetic engineering, but the entire relationship between biotechnology and our future. He concludes that it is important to not lose sight of the larger policy questions that will determine whether our ability to manipulate the basic process of life will benefit the world community.

Social science research can enhance the debate by providing a systematic and credible mechanism for incorporating societal values and preferences into public and private decisions.

Social science research can enhance the debate by providing a systematic and credible mechanism for incorporating societal values and preferences into public and private decisions. Social science provides useful insights for the important policy debates and educational programs that are needed. However, social science often tries to reach a moving target. Public perceptions are complex and dynamic. It is important to recognize that many factors will ultimately affect consumer acceptance of foods developed through biotechnology. Public awareness and attitudes will change as more information and actual food products become available. In that sense, any research project serves as a baseline for future work.

PROJECT METHODOLOGY

The purpose of the project was to determine what consumers think and want to know about the use of biotechnology in food production and processing. Results should prove useful in several ways. One outcome will be to provide guidelines for development and implementation of educational programs. Another outcome will be to enlighten the formulation and evaluation of public policies by providing a valid means of citizen input. Third, the results will recommend appropriate and acceptable types of biotechnology research and product development activities.

This project involved several complementary research methodologies. All work was conducted in a collaborative manner using established social science methods. Full details on project methodology can be found in the technical reports which are available from the authors (Hoban and Kendall, 1992; Kendall and Hoban, 1993). An extensive literature review assessed existing studies on public attitudes about biotechnology, as well as surveys on similar topics. From this review, conceptual models were developed that guided survey development.

Several organizational structures were developed to carry out this project in a consultative and interdisciplinary manner. At North Carolina State University and Colorado State University, interdisciplinary project teams were established. In addition, a national advisory committee consisting of university and government experts was established. The campus teams and national committee assisted with development and review of the telephone survey and focus group instruments, as well as the technical reports.

Third, the results will recommend appropriate and acceptable types of biotechnology research and product development activities.

Development of the telephone survey instrument involved several related tasks. At the outset, a series of open–ended individual and small group interviews were conducted in Washington, DC with public and private sector leaders in the area of biotechnology. Results of the interviews were summarized to inform development of the telephone survey. To further inform survey development and test the face validity of potential questions, personal interviews were conducted with about 40 consumers in North Carolina and focus groups were conducted with 67 consumers in Colorado, Ohio, New York, Pennsylvania and Nebraska. Results of these interviews and focus groups led to further refinement of the telephone survey instrument. Additional telephone pretests were then conducted to finalize question wording.

The sampling frame for the telephone survey was the entire population of households in the United States with telephones. The random sample of telephone numbers was purchased from Survey Sampling Incorporated (SSI). Their samples are systematically drawn using random digit dialing. Both listed and unlisted numbers are included in the sample. All telephone interviews were conducted by the Applied Research Group at North Carolina State University during February and March of 1992. A total of 1,228 interviews were completed. The final telephone interviews averaged about 26 minutes in length. At least ten attempts were made at various times to reach each telephone number selected before a number was eliminated from the sample. The response rate for the survey was just over 60 percent. The sample population appears quite representative of the country as a whole on most major background characteristics. Full demographic profiles of respondents are included in the project technical report (Hoban and Kendall, 1992).

After the telephone survey, focus groups were conducted during October of 1992 to assess reactions to specific food products, obtain more detailed opinions about biotechnology, and have consumers generate ideas about educational needs. Subjects were randomly selected from city telephone directories in Denver and Raleigh. Eight focus group sessions were held: four in Colorado and four in North Carolina. A total of 46 people participated in the eight groups. The focus–group lasted approximately two hours. Discussion focused on eight topics: general knowledge of biotechnology; reaction to specific products; information needs and sources; labeling issues; environmental issues; economic issues; moral and ethical issues; and the role of citizens. In general, focus groups results supported many of the phone survey results and added additional qualitative insights. Full details can be found in the Focus Group Technical Report which is available from the authors (Kendall and Hoban, 1993).

SELECTED SURVEY RESULTS

This paper focuses on selected results of the telephone interviews. In keeping with the NABC 5 theme, we mainly discuss findings related to public perceptions of the benefits and risks of biotechnology. Some key relationships between a selected set of background characteristics and general reaction to biotechnology will be analyzed. Due to limits on length, we cannot present information on all the background variables. Full information, including the wording for all questions and detailed analysis results, are included in the Telephone Survey technical report (Hoban and Kendall, 1992).

GENERAL ATTITUDES ABOUT BIOTECHNOLOGY

Respondents were read several statements about the general benefits and risks of biotechnology and asked the extent to which they agreed or disagreed. Almost three quarters agreed that "Biotechnology will personally benefit people like me in the next five years." On a related statement, over two–thirds agreed that "Government should fund more biotechnology research because of the potential benefits." Concerns about risk are evident by the fact that nearly half of the respondents agreed that "Biotechnology should not be used because of potential risks to the environment."

...most people did not know enough about biotechnology to provide a detailed assessment of potential benefits or risks.

As we designed the survey, it became evident that most people did not know enough about biotechnology to provide a detailed assessment of potential benefits or risks. One set of questions did measure respondents' general expectations concerning the potential impacts of biotechnology. Overall, respondents anticipated generally positive effects of biotechnology in most areas. Almost three quarters of the respondents saw the effects as positive for farmers' economic conditions, as well as for food quality and nutrition. Nearly two–thirds felt the effects of biotechnology would be positive for environmental quality. Between one–half and two–thirds of the respondents saw positive effects of biotechnology on farmers' use of chemicals and fish and wildlife. In all cases, between five and ten percent of all respondents did not have an opinion about whether the effects would be positive or negative.

One pair of questions assessed whether respondents would have moral objections to the use of biotechnology in either animal or plant applications. The first question asked: "Do you believe the use of biotechnology to change plants is morally wrong or not?" In this case, almost one quarter of the respondents felt that it would be morally wrong. Respondents had stronger views concerning the moral aspects of animal biotechnology. When asked "Do you believe the use of biotechnology to change animals is morally wrong or not?" over half said that it was. This turns out to be one of the most important and unique issues associated with the use of biotechnology in agriculture and food production.

This [morality] turns out to be one of the most important and unique issues associated with the use of biotechnology in agriculture and food production.

Acceptance of Biotechnology

The interview attempted to address a fairly complex and controversial area involving consumer acceptance of food products that involved clearly transgenic characteristics. Such issues have received considerable attention from the media, consumer groups and industry. Examples were used that either reflected actual applications or were representative of possible future uses. These questions were sequenced to move from the least dramatic to most dramatic examples. This set of questions was introduced with the following statement: "Genes from most types of organisms are interchangeable." Respondents were first asked "Would potatoes made more nutritious through biotechnology be acceptable or unacceptable to you if genes were added from another type of plant, such as corn?" Two–thirds of all respondents said they would find such plant–to–plant gene transfer acceptable. Respondents were next asked "Would such potatoes be acceptable or unacceptable to you if the new genes came from an animal?" In this case, only one quarter of all respondents indicated they would find such animal–to–plant gene transfer acceptable.

Two examples were used to determine reaction to animal–related gene transfers. First, respondents were asked: "Would chicken made less fatty through biotechnology be acceptable or unacceptable if genes were added to the chicken from another type of animal?" In this case, just under 40 percent of the respondents said they would find such a gene transfer acceptable. As a final, relatively dramatic (but technically feasible) application, respondents were asked: "Would such chicken be acceptable or unacceptable if the genes came from a human?" Only 10 percent of all respondents indicated that such human–animal gene transfer would be acceptable.

Public Policy and Citizen Participation

One set of questions examined the area of government credibility. Respondents were asked if they would have a lot, some, or no confidence in the U.S. Department of Agriculture (USDA), Food and Drug Administration (FDA), Environmental Protection Agency (EPA), and state agencies to effectively regulate biotechnology. In all cases, about two–thirds of the respondents reported "some" confidence. Of the remaining respondents, the percent reporting "no confidence" exceeded the percent reporting "a lot" of confidence. Differences among the levels of confidence in each of the agencies were quite small.

There seems to be considerable interest in biotechnology.

Two statements measured respondents' opinions about the role of citizens in biotechnology decisions. Over three quarters agreed that "Citizens have too little say in decisions about whether or not to use biotechnology." On a related point, almost all respondents agreed that: "Government should pay more attention to what people like me think about biotechnology." This likely reflects more general feelings that people have about government responsiveness and effectiveness.

Interest in Biotechnology

There seems to be considerable interest in biotechnology. This suggests that biotechnology will be a major public policy issue. One in five respondents said they had a lot of interest in learning more about biotechnology. Almost half reported some interest. One in five had only a little interest and the remaining 14 percent said they had no interest in learning more about biotechnology.

Those respondents who reported at least "a little" interest in learning more about biotechnology were then asked how important they felt it would be for them to receive each of six different types of information regarding biotechnology. This indicates the relative level of importance people attach to each of the major issue areas. Over two–thirds felt it would be "very important" for them to have information about the potential risks or negative effects of biotechnology. Most of the rest said such information would be somewhat important. Just under two–thirds said it would be very important to have information about new uses of biotechnology in human health care. About half indicated that it would be very important to have more information about the potential benefits or positive effects of biotechnology. Under half felt information about new uses of biotechnology in food production would be very important. Almost as many felt that information about how government regulates biotechnology would be very important. The information considered least important involved the basic science behind biotechnology. However, even in this case over one–third said this information would be very important and over half said it would be somewhat important.

Overall Reaction to Biotechnology

It seemed important to obtain a general assessment of respondents' bottom–line reaction to biotechnology. This was done by asking: "Overall, would you say you support or oppose the use of biotechnology in agriculture and food production?" Almost two–thirds of all respondents said they supported the use of biotechnology in agriculture and food production. Just over a quarter opposed its use. Almost 10 percent did not have an opinion on the question.

Some additional analysis reveals the types of respondents who were likely to support the use of biotechnology in agriculture and food production. In terms of demographic and background characteristics, men were more likely to support the use of biotechnology than were women. Respondents with higher educational and/or income levels also expressed greater support for biotechnology. People who considered religion to be more important in their daily lives tended to express greater opposition to the use of biotechnology. Interest in new scientific and technological developments was positively correlated with support for the use of biotechnology.

Interest in new scientific and technological developments was positively correlated with support for the use of biotechnology.

Other attitudes about and awareness of biotechnology were also significantly related to support for the use of biotechnology in agriculture and food production. Those who had read or heard more about biotechnology, as well as respondents with a greater interest in learning about biotechnology, expressed greater support. Respondents who felt biotechnology was morally wrong were much more likely to oppose its use. High levels of confidence in government, as well as trust in information, had a significant positive relationship with support for biotechnology.

Reasons for Supporting Biotechnology

About two–thirds of all respondents answered positively when asked whether they supported or opposed the use of biotechnology in agriculture and food production. A wide range of answers were given in response to the follow–up question: "Can you describe why you support the use of biotechnology?" Seven major categories were developed that describe the reasons. Open–ended questions like this are useful for gaining deeper insights into peoples' perceptions.

The most common reason for support of biotechnology (represented by 22 percent of all responses) was that people believed it will enhance the quality of food products. Many of these respondents indicated that by using biotechnology the nutritional value, taste or other desirable qualities of food could be improved. Some suggested that biotechnology could be used to reduce fat, cholesterol, pesticide residues or other undesirable aspects of food products. Others indicated that biotechnology would make food safer.

The most common reason for support of biotechnology ... was that people believed it will enhance the quality of food products.

Of those who supported biotechnology, about one in five believed it would help increase the quantity of food produced. Many of those in this group expressed the opinion that food production needs to keep up with population growth. Some respondents viewed biotechnology as a means of reducing hunger. Biotechnology was considered by some to be an important method for producing more food at lower cost, with the use of less land or other resources.

Approximately 14 percent of supporters felt biotechnology will benefit society by providing environmental or economic benefits. Some respondents in this category indicated that biotechnology will be beneficial to American industry and reduce the cost of production. Other individuals indicated that biotechnology may help reduce pollution, including agriculture's dependency on chemicals. Some respondents specifically suggested that farmers would benefit from the development of biotechnology.

Respondents... basically felt new developments in science and technology were necessary and desirable.

Another group of respondents (14 percent) suggested that scientific and technological advances benefit society, in general. Respondents viewed biotechnology as important for maintaining such leadership in the future. Respondents in this category basically felt that new developments in science and technology were necessary and desirable. Approximately 12 percent provided statements which suggested biotechnology will improve the overall quality of life. More specifically, some respondents in this category hoped it would improve health care, including finding cures for medical problems. Ten percent of the responses involved nonspecific statements of support for biotechnology. These responses simply implied a belief that biotechnology is good or beneficial without being able to provide specific reasons.

Nine percent of the respondents who supported biotechnology gave statements that reflected some apprehension. Some indicated there may be problems with biotechnology. Other respondents said they worried whether the experts would consider citizens' best interests. Some responses indicated a willingness to support some applications (e.g., with plants) but not others (e.g., with animals). Overall, individuals in this final category indicated their support for biotechnology was conditional.

Reasons for Opposing Biotechnology

Just over one–third of all respondents who took a position on this question opposed the use of biotechnology. They were then asked: "Can you describe why you oppose the use of biotechnology?" Responses were again coded into several main categories. The most common reason for opposition to biotechnology (given by over one–third of all respondents who opposed biotechnology) involved concerns that it could threaten the balance of nature. Some respondents worried that biotechnology would lead to serious impacts on the natural environment. They felt such tampering with nature was wrong. Other respondents in this category opposed biotechnology because it was "not natural." Some mentioned concerns about loss of genetic diversity or the creation of harmful mutations.

The most common reason for opposition to biotechnology ...involved concerns that it could threaten the balance of nature.

Around thirty percent of those who opposed biotechnology said they were concerned about its unknown effects or long–term risks. Some respondents in this category mentioned lack of trust in scientists or government to adequately control biotechnology. Others felt it could be used in an inappropriate manner. Some felt that not enough testing would be done regarding the possible side effects. Such reasons seemed to revolve around the general notion of perceived risk.

The third most common reason (about 13 percent) for opposition involved concerns over the application of biotechnology with either humans or animals. Respondents also worried about possible impacts on human population growth. About eight percent of the respondents who opposed the use of biotechnology had concerns that it would promote the increased use of chemicals in farming or food processing. Some worried that food safety or quality would be negatively affected through the use of biotechnology, resulting in possible harm to those who eat the food products.

Another eight percent of the comments from those opposing biotechnology indicated that respondents opposed the use of biotechnology because they believed its was somehow against God's will or contrary to their religious beliefs. This reason also included a general sense of moral objection, as well as objection to trying to improve on "God's creation." Four percent of those who opposed biotechnology did so because of concern about impacts on the economy. Individuals mentioned potential social or economic impacts that biotechnology could have for small businesses or family farms. Two percent of the statements involved concerns over other impacts.

A significant commitment to unbiased and ongoing education is needed.

CONCLUSION AND IMPLICATIONS

This research project has a variety of implications for the development and implementation of educational efforts and public policies. In this final section, we describe some of the main implications in these areas. This section will conclude with a discussion of future social science research needs, including the limitations of such work.

Implications for Education and Policy

Results of this work document the need for expanded education and information efforts. Given relatively low levels of awareness and considerable interest in biotechnology, a significant commitment to education is needed. Outreach programs need to be developed and implemented to provide people with information they need to better understand the issues related to biotechnology. The goal should be to help consumers make their own informed decisions about this new technology. This should involve a broad–based approach aimed at school children, organizations, opinion leaders and consumers. A wide range of groups and organizations can contribute to educational programs. Land–grant universities, such as those represented by NABC, are in a credible and influential position to play an important leadership role in such education.

The goal should be to help consumers make their own informed decisions about this new technology.

This project indicates that consumer knowledge and attitudes about food produced through biotechnology will be influenced by general information obtained from the media. Education efforts must recognize the limitations and opportunities for media–based information. Education needs to be unbiased, ongoing and proactive. Adequate time, money and expertise must be devoted to education to ensure that opinion leaders and consumers are able to receive the information they want and deserve in a timely and credible manner. The future of biotechnology in food production is by no means assured without a much more proactive and open dialogue among all interested parties.

Our results indicate that one of the most important factors influencing public perception of biotechnology will be the perceived credibility of public policies and regulations. Consumers want government to play an active role in establishing policies that ensure environmental protection and food safety. They also want government to expand the debate over the most appropriate uses of this powerful set of technologies. The federal government should pursue a proactive and credible approach to biotechnology policy that empowers citizens to make informed decisions, while facilitating development of appropriate products.

...public confidence in government agencies to effectively regulate biotechnology could be significantly improved.

This project indicates that public confidence in government agencies to effectively regulate biotechnology could be significantly improved. Few people understand the nature of government policies and regulations in this or other areas. Low confidence is, in part, a reflection of a more general erosion of public confidence and trust in government. Attempts during the past decade to reduce regulation have been responsible, in part, for decreased public confidence in government. Most respondents felt that government should pay more attention to what citizens think about biotechnology. People want more say in decisions that affect their lives. This is not unique to biotechnology, but includes other policy areas. Consumers expect public policies to be developed in an open manner with ample opportunities for involvement of all interested stakeholders.

Future Research Needs

Biotechnology is a complex and dynamic public policy area. This project may have raised as many questions as it has answered about public perception of biotechnology. Because the telephone survey was done before the May, 1992 FDA announcement of regulations for food produced using biotechnology, it represents a baseline for future surveys using the same questions. This final section will offer some general observations about the utility and limitations of social science research.

Public attitude surveys, such as this one, are useful for identifying issues, interests, concerns, educational needs and public policy implications. Our work provides insights into these areas. For example, this work shows different levels of consumer acceptance for different products. It also highlights important ethical and environmental issues. Our results also suggest appropriate education and information efforts.

However, telephone surveys are not good for evaluating specific products or probing deeply into people's basic beliefs. Other methodologies will be more appropriate and useful for testing consumer response to new food products. Focus groups and taste tests will provide guidance for marketing of specific products. Market–basket studies and computer simulation models will help evaluate and predict actual consumer behavior. Once people are actually able to taste a new food product, they will form much more definite opinions about its desirability.

Survey research is as much an art as a science. It is important to start with a clear set of objectives and a testable conceptual model. Research must be based on past research and theory as well as clear plans for analysis. Social science research is most useful when it goes beyond simple description into the realm of explanation and prediction. It is necessary to analyze why people feel the way they do. The population must be segmented into different groups and compared on key questions. Our analysis along these lines is just beginning. Further results will be published in professional journals.

Realize that survey research is only one piece of the puzzle. Regulations must still be based on the best available science. Public policy will always include economic and political considerations. Surveys, however, do provide a cost–effective and systematic mechanism to incorporate public beliefs and values in decisions. If done right, surveys can reflect the views of a diverse cross section of citizens.

ACKNOWLEDGMENTS

We received assistance and advice from many knowledgeable people representing various government agencies and universities. Funding was provided by the Extension Service of the U.S. Department of Agriculture, North Carolina State University, and Colorado State University. The conclusions presented here do not necessarily reflect the views of the funders or others affiliated with this project.

REFERENCES

Cross, H. R. 1992. Food Safety Perspectives on Animal Biotechnology. In *NABC Report 4, Animal Biotechnology: Opportunities and Challenges.* J. Fessenden MacDonald, ed. National Agricultural Biotechnology Council, Ithaca, NY. p. 121–126.

Foreman, C.T. 1990. Food Safety and Quality for the Consumer: Policies and Communication. In *NABC Report 2, Agricultural Biotechnology, Food Safety, and Nutritional Quality for the Consumer.* J. Fessenden MacDonald, ed. National Agricultural Biotechnology Council, Ithaca, NY. p. 74–81.

Gore, A. 1991. Planning a New Biotechnology Policy. *Harvard Journal of Law and Technology.* 5(Fall):19–30.

Hoban, T. and P. Kendall. 1992. *Consumer Attitudes About the Use of Biotechnology in Agriculture and Food Production.* North Carolina State University, Raleigh, NC.

Kendall, P. and T. Hoban. 1993. *Consumer Attitudes about Food Produced with Biotechnology: Moral Opinion/Gender Focus Groups Technical Report.* Colorado State University, Ft. Collins, CO.

Office of Technology Assessment (OTA). 1992. *A New Technological Era for American Agriculture (Summary Report).* U.S. Government Printing Office, Washington, DC.

Stenholm, C.W. and D.B. Waggoner. 1992. Public Policy in Animal Biotechnology in the 1990s: Challenges and Opportunities. In *Animal Biotechnology: Opportunities and Challenges.* J. Fessenden MacDonald, ed. National Agricultural Biotechnology Council, Ithaca, NY. p. 25–35.

Stiles, S. 1989. Opinion. In *Biotechnology Decisionmaking: Perspectives on the Objectives of Public Participation.* The Keystone Center, Keystone, CO. p. 7–9.

Public Values and Risk Assessment

Roger A. Balk
Ethicist,
McGill University;
Coordinator,
Physician's Information Systems,
Royal Victoria Hospital,
Montreal

I want to begin by discussing Henny Penny.[1] To the best of my knowledge, she has not been patented nor is she a registered trademark, but as is appropriate for my presentation, she is eminently in the public domain. This story may be one of the first encounters most of us have had with the values relevant to risk assessment and since most public presentations about risk are in the form of stories, albeit television or newspaper, it may help to remind ourselves how well this childhood pastoral legacy fits present day circumstances.

You know the story. Henny Penny is in the barnyard when suddenly she is hit on the head by an acorn. She immediately assumes the sky is falling and that she must hurry to tell those in authority. She recruits a number of her companions to join in the mission. On the way she and her friends are seduced by a wily fox to take a short cut from which she and her friends are never heard from again.

It presents a role model which derides conclusions based upon a foolish reaction to Nature because this response leads to even greater disaster. Nature is very regular; she does not play tricks such as sending the sky to fall like rain and we can depend on that when we try to figure out what is going

[1] Presently there are 10 listings for this story in Canada and 8 in the United States in current *Books in Print.*

on around us. True natural events which can be very destructive are not ruled out, but we can deal with these precisely because they are a part of common sense understanding of what the world is about. We must respect and can trust nature in its untampered state.

The presence of technology has altered this picture. Based on science, it tries to harness nature's regularity and tame her power to produce predictable outcomes upon which a host of human activities can be formed. On this view, animal domestication may be one of the earliest examples of a technological change being imposed upon the natural world. Bronowsky (1973) has argued that the discovery of a cultivatable form of wheat is one of the primal discoveries of civilization. But there is a slight caveat to which I would like to call your attention. Most people do not view the domestication of animals or the discovery of wheat as technology. They see it as an example of human ability to exploit natural abundance, fundamental to the Story of Creation which gives biblical authorization to pastoral goodness. A contemporary example of how pervasive this attitude is, is found in Michael Fox's introduction to his book, *Superpigs and Wondercorn: The Brave New World of Biotechnology...and Where It All May Lead* (1992). In it he describes the need to repair "this dispoiled planet" and the need to "dress and keep" the Garden of Eden.

What for the average North American resident symbolizes the state of contemporary agriculture as well as what to expect from biotechnology consists of a brew of naiveté and skepticism—Arcadia or the Monster.

The views just presented are not easily reconcilable. They offer conflicting approaches to the course of human life. Most Americans know little of the history of science and technology. What for the average North American resident symbolizes the state of contemporary agriculture as well as what to expect from biotechnology consists of a brew of naiveté and skepticism—Arcadia or the Monster. It is not hard to find vestiges of this style of moral understanding in current conflicts over the licensing of agricultural products having biotechnological modifications.

Here are three illustrations which I believe represent current versions of this state of Americana: First, there is the belief that if you grow something, it represents contact with a reality absolutely fundamental for human existence. Apartment dwellers with their three tomato plants 30 stories above the street can really get into this business of growing things. (Of course, it may also be a last desperate attempt to find a replacement for the wooden tomatoes which adorn the average salad.) At this level, these plants can signify a desire to maintain one of the last vestiges of the pastoral dream from which most of life has been wrenched—roots, if you will.

Second, there is the paradox inherent in the abundance which North American agriculture exemplifies. On the one hand, the industry remains one of the major unsolvable problems for modern governments, far outstripping the complications currently posed by the challenge to transform our defense industry. On the other hand, this politicalization of agriculture has succeeded in disillusioning the very people who wallow in its largess. The difference in the effectiveness of the pork barrel in determining agricultural policy, while at the same time the failure of legislators to protect the consuming public from the risks inherent in the raising and preparation of food for consumption, has not gone unnoticed, *viz* federal inspection of the meat packing industry.

Third, the effects of biotechnology as it applies to agriculture are a source of concern both rational and irrational. There is a real ignorance of science and how it works especially as related to probability. There is a belief that the effects of biotechnological manipulation pose a risk for a possible but unknown catastrophe. To the extent that the changes biotechnology proposes initiate an element of risk to those who use its products, they demand a form of control unknown to the simple morality of a pastoral ideal in conflict with an apocalyptic vision of a mechanical universe which would destroy the Garden. To the extent that the present effectiveness of the protective role of government has a high failure rate with no attribution of responsibility, there is real fear as to what Food and Drug Administration (FDA) approval means in terms of protecting the consumer. This view of the failure to protect from risk can be dismissed as the continual failure of moral politics to understand and regulate the new face of agriculture. Finally, the extent to which contemporary patterns of food distribution make it difficult for the average person to chose on the basis of accurate information—increases the paranoia of a public ignorant of science and fuels the notion that current regulatory procedures controlling agricultural biotechnology are untrustworthy.

...the effects of biotechnology as it applies to agriculture are a source of concern both rational and irrational.

That Americans were far from being prepared for a change of moral climate was suggested in 1964 by Leo Marx in his monograph, *The Machine in the Garden.* His review of American literature from the colonial period to the present day suggests that we are stunned by the magnitude of the protean conflict figured by the machine's increasing domination of the visible world. This recurring metaphor of contradiction makes vivid, as no other figure does, the bearing of public events upon private lives. It discloses that our inherited symbols of order and beauty have been divested of meaning. It compels us to recognize that the aspirations once represented by the symbol of an ideal landscape have not, and probably cannot be, embodied in our traditional institutions. It means that an inspiriting vision of a humane community has been reduced to a token of individual survival. To change the situation we require new symbols of possibility, and although the creation of these symbols is in some measure the responsibility of artists, it is in greater measure the responsibility of society. The machine's sudden entrance into the garden presents a problem that ultimately belongs not to art, but to politics (Marx, 1964).

CLUES FROM BIOETHICS: INFORMED CONSENT

As I see it, the challenge presented by the new age is to transform our simplistic view of moral conflict with a nuanced theory of ethical accounting. It requires vocabulary which reflects awareness of the content of the public values just presented and a theoretical structure that incorporates the reason of science into our political reality. The paradigmatic instance of this change is played out in the history of bioethics over the last quarter century. Now a mature enterprise, lessons from what has happened in medicine may provide some clues as to how traditional moral responses to biotechnology might be recast in a mold which will resolve current social deadlock.

...the challenge presented by the new age is to transform our simplistic view of moral conflict with a nuanced theory of ethical accounting.

The history of bioethics has been covered elsewhere (Clouser, 1970). For our purposes, I would emphasize its source in the awareness that *individuals* have the *right* to be protected from exploitation by those who are offering them medical care and treatment. Over the years, a series of cases have helped delineate the circumstances under which this protection should apply (NIH, 1980). The result has been the creation of a working system whereby technology is supported at the same time protecting those who would be its beneficiaries or its victims (e.g., Halushka vs. University of Saskatchewan, 1965). In the center of this system has been FDA. There are three tiers to this system which are indispensible to its functionality: Tier One—the establishment of ethical principles which must be realized in action; Tier Two—the demand that all activity must be backed by data which has been subject to the statistical demands demanded by contemporary science; and Tier Three—the development of a process whereby the first two criteria may be acted upon. In addition, and indispensable to the working of the system, is a definition of roles which separate the function of regulatory as opposed to developmental responsibilities. Let me describe in more detail the constituents of this practice:

Tier One

The establishment of ethical principles which provide a moral arena for action. Here the dominant normative force has been autonomy. Functionally, this has led to the enshrinement of informed consent as its most important expression. Subject to a caveat covering emergency treatment for children and the mentally incompetent, every act with its costs and benefits must first be understood before being accepted by an individual patient. The extension of risk to patients other than mentally competent adults is severely restricted to those interventions for which the benefit is clearly demonstrable. As it has developed, the practice of principled action has been most completely explored as it applies to individuals. Principles governing society, such as justice and fairness, have only recently been the subject of more sophisticated attention, particularly as it applies to the availability of health care. Here it is obvious that notions such as that of "the common good" are highly controversial and the sharing of both costs and benefits has proved highly difficult to implement in a socially acceptable manner.

Tier Two

The requirement that all projected actions must be the subject of investigations which produce data conforming to current scientific practice as regards statistical probability is essential to developing a meaningful cost/benefit analysis. Of more than a little interest to those concerned about the effects of biotechnology are the practices which cover the development and use of new drugs. The three–phase trial system which moves from animals to human subjects is used to determine general parameters of risk and efficacy (NIH, 1977). Only after passing all three phases can a drug be licensed for the task for which it was tested. This system is not perfect, but its shortcomings have not prevented it from working. Of some concern is the practice of using only adult males in the Phase 3 trials, the final step before licensing. The failure to include children, women and the elderly in these protocols has led many to question the conclusions, particularly about dosage and side effects which are included in the approval documentation. The practice of asking physicians to report adverse effects as they occur in the "field" has only been partially successful in developing a more complete dossier about each drug.

Tier Three

The development of a process through which principles may be combined with data to produce a distribution system which is safe within defined limits acceptable to the consuming public is the final component in the practice Institutional Review Boards (IRBs) must review all trials using human subjects for both scientific merit and ethical responsibility (Levine, 1961). The record has been neither incident nor scandal free.[2] There was the famous thalidomide affair and perhaps more to our concern, the cancer risks to the daughters of mothers who had been prescribed diethylstilbestrol (DES) during

---

[2] The most outspoken of recent critics has probably been Ivan Illich. c.f. *Limits to Medicine: Medical Nemesis: The Expropriation of Health.* McClelland and Stewart, Toronto, Ont. 1976.

pregnancy as an attempt to avoid early miscarriages (Potter, 1991). There have been detractors who insist that the present system is far too conservative in a time of crisis such as that produced by AIDS. Nonetheless, it is critical not to lose sight of the core process. It is the principle of informed consent. This has allowed for the implementation of such widely differing practices as giving bioengineered growth hormone to children, gene therapy for cancer, and xenografts from hogs with transgenically altered immune systems. Each individual can in theory and in practice, supported by social consensus, assume risk because each is free to chose whether or not to receive defined benefits. Consent is also understood as a process with several gradations—with increase risk in relation to benefit requiring more awareness of what is being accepted.

The additional element in the practice just described is what, I believe, accounts for its level of public acceptance. This is the attempt to separate regulatory responsibilities from developmental functions—the role of government as distinct from that of industry. Jane Jacobs in her recent book, *Systems of Survival–a Dialogue on the Moral Foundations of Commerce and Politics* (1992), argues that these groups represent two distinct modes or syndromes, the Guardian and the Commercial, which are essential to the functioning of human society. The first, guardianship, arises from the behavior which we share with animals—foraging for food and protecting our territories. Guardians work in the armed forces and police, government ministries and their bureaucracies, legislatures, courts and organized religions. The second, Commercial, arises from trade and production of goods and is an endeavor unique to human beings. These two modes of survival have produced two discrete and contradictory ethical systems and are the source of conflict when the precepts appropriate to the guardian system are imposed on the commercial and vice versa. In its everyday functionality, this means that drug companies are free to be as inventive as possible—expressed as profitability—so long as they operate within the structures and regulations designed to protect society. But, and this is a major qualifier, the individual (a patient in this case) is still free to determine whether he or she accepts the risks and benefits made available by this symbiotic structure. In the case of health care, the point where this assumption of risk and benefits occurs is not in terms of a market relationship, but as informed consent. As we turn to the matter of agricultural biotechnology, we encounter a significant difference in the risk/benefit struc-

There is great difficulty in exercising informed consent [in agricultural biotechnology] because...the monolithic distribution system tends to restrict action to all or nothing.

ture. There is great difficulty in exercising informed consent because, as we have previously noted, the monolithic distribution system tends to restrict action to all or nothing. You either buy the product available or you do without it. In practical terms, consumers are left with what appears to be an irrational response—massive group threats of boycott—to what is more reasonably viewed as a need for rational discussion and understanding.[3]

CURRENT CONTROVERSIES: AN ETHICAL ANALYSIS

In this final section I would like to review a number of current controversies involving bioengineered agricultural applications, subjecting them to the three–tiered structure I have proposed and which has been developed in the course of bioethics as well as the survival ethics proposed by Jane Jacobs (1992).

I will begin with the Tier Two of the structure. This is the requirement that all projected actions must be the subject of scientific investigation. I think that what is most upsetting to the scientific community is the fact that on this criteria, agricultural biotechnology has performed quite well. Let me begin by comparing the cases of bovine and porcine somatatropin. The evidence would suggest that, from the point of animal welfare, bST is acceptable—the review of testing seeming to indicate that there is no increased risk of mastitis in animals given the hormone, as well as attesting that there is no contamination of the milk by bST. The effect upon hogs by pST has not been so benign and until the adverse effects on the animals can be controlled, on the basis of animal welfare alone, scientific evidence would tend to support withholding acceptance of this method of enhancing lean qualities of pork. In both of these examples, there does not appear to be serious objection to *in vivo* investigations. When it comes to *Bt*–toxin in food crops to control insects, the issue is more complex. Recent field tests of corn involving transgenic manipulation of an insecticidal protein derived from *Bacillus thuringiensis (Bt)* has proved it to be highly resistant to heavy field infestation of the European corn borer (Koziel et al., 1993). Here, I believe, our experience from medical history may be helpful in delineating the issues. Are the long–term effects of *Bt*–toxin similar to what has happened to the effectiveness of our recent treatments for tuberculosis or to the eradication of small pox? Or to put the problem in another way, what is the evidence that the long–term effects of *Bt*–toxin on pest control will be more successful than produced by the heavy use of insecticides? Then, there is the controversy of the Flavr Savr™ tomato. The likelihood that its use of the kanamycin–resistance gene

[3] Compare the campaign waged by Jeremy Rifkin against the Flavr Savr™ tomato which involved threated boycott of McDonald's and Campbell's products as well as the enlistment of prominent chefs to refuse their use.

as a marker in the reverse–RNA antisense process will produce human side effects by hooking up with the wrong bacteria during the digestive process is highly remote (Hoyle, 1992). The evidence appears to be at a level of certainty which would be perfectly acceptable had this been a drug being considered for licensing. It is not an overstatement to suggest that the level of scientific research and testing involved in biotechnology is at least as good as that available in the health care field. If we want to understand what may be the problem, we must turn to the other tiers which round out the consumer protection package available to the user of health care.

How can government regulate with one hand when its other hand is in the business of agriculture?

This brings us to another look at Tier One—the establishment and implementation of ethical principles to provide a normative element that protects the user/consumer from exploitation. I do not believe that different principles apply here from those which are used in our evaluation of health care. But our previous examination does suggest that informed consent by the individual to the risk/benefit involved in food product use has a more complex ethical application.[4] Furthermore; the nature of risk is such that the specific instance in which it will appear cannot be determined, so that concepts of common good may unwittingly but unfairly single out victims. Theoretically it could be possible to establish a compensation system to help ease the morbidity/mortality of victims, but the multisource of present risk for disease such as cancer would make the application of this worthy idea almost impossible. So, I believe the resolution of our ethical case must take into account the general insistence of the American consumer that the decision to assume risk must be an individual one, even if there are notable instances when the actual rational weighing of outcome is honored more in the breach as, for example, in deciding to get a driver's license.

Can the process powers of Tier Three produce a distribution system which is safe and accountable by reworking the ways in which we apply our ethical principles? We may get a better sense of what this orientation is up against by asking some leading questions. Is McDonald's likely to give customers a choice of Flavr Savr™ tomatoes or regular ones? What about Campbell's? Labeling is one possible response, but it is ethically acceptable as reflecting the existence of choice only if alternatives are readily available. Merely spelling out contents is not enough because what is of paramount im-

---

[4] For a different approach see: MacLean, D. "Social Values and the Distribution of Risk." In *Values at Risk.* D. MacLean, ed. Rowman & Allanheld, Totawa, NJ. 1986. p 75-93.

portance is some indication of risk. Is there the equivalent available to us of a Phase 3 drug trial which could produce acceptable accounting of risk? (It would of course add considerably to the cost of food in the short run.) Who could run a large trial of 10,000 to 20,000 randomized participants? For what period of time? Most carcinogens are notoriously slow in producing symptoms of disease so we would probably have to accept animal evidence here. Possibly a consensus conference, something like the Presidential Commission on Bioethics (1980) that produced the original ethical guidelines for health care could, in addition to offering insight as to the relationship of autonomy to common good, clarify the roles of government and industry. For example, only government in the exercise of its guardianship role could extend the common good to include the ecosystem and produce the regulations to which development must conform. Certainly we need less of the kind of argument presented by Michael W. Fox (as cited by Johnson, 1993) that implies that eliminating profit motives is in the public interest, because it confuses even further the difference between protection and innovation, both of which are vital to our future welfare. This discussion has also shown that the social interaction we have called "process" is all too often subsumed under the term *politics.* Nowhere is this more true than in that paradoxical enterprise of our society called agriculture.

...the general insistence of the American consumer [is] that the decision to assume risk must be an individual one...

Jane Jacobs (1992) offers the vision that the human past and future is tied in quite absolute ways to the proper use of both guardian and commercial enterprise. The application of biotechnology to agriculture seems destined for more of the same unproductive confusion and mistrust by our citizenry unless we can sort out the current confusion as to which is responsible for what. How can government regulate with one hand when its other hand is in the business of agriculture? Since the time frame of all living creatures is from the human perspective, is it nonetheless a fitting human response to expect that profitability take a somewhat longer perspective than the next two to five years? I would offer, as one possibility, that the combination of values which currently drives American skepticism about the future promised by biotechnology is demanding a standard of accounting closer to a view of time expressed by the evolution of natural life than the short term perspective that plagues both government and industry. *Sub specie aeternitatis* indeed!

ACKNOWLEDGMENT
My thanks to Kris K. Carter D.V.M. for her helpful comments on earlier versions of the manuscript.

REFERENCES

Bronowsky, J. 1973. *The Ascent of Man.* Little Brown and Co., Boston, MA. p. 64.

Clouser, K.D. 1978. Bioethics. In *Encyclopedia of Bioethics.* Free Press, New York, NY.

Fox, M.W. 1992. *Superpigs and Wondercorn: The Brave New World of Biotechnology...and Where It May All Lead.* Lyons and Burford, New York, NY.

Hoyle, R. 1992. Winning the Tomato War. *Bio/Technology.* 10(Dec):1520–1521.

Jacobs, J. 1992. *Systems of Survival & a Dialogue on the Moral Foundations of Commerce and Politics.* Random House, New York, NY.

Johnson, R. 1993. Book Review (citing M.W. Fox in *Superpigs and Wondercorn*). *Bio/Technology.* 11(Feb):180–181.

Koziel, M., G. Beland, C. Bowman, N. Carozzi, R. Crenshaw, L. Crossland, J. Dawson, N. Desai, M. Hill, S. Kadwell, K. Launis, K. Lewis, D. Maddox, K. McPherson, M. Meghji, E. Merlin, R. Rhodes, G. Warren, M. Wright, and S. Evola. 1993. Field Performance of Elite Transgenic Maize Plants Expressing an Insecticidal Protein Derived from *Bacillus thuringiensis. Bio/Technology.* 11(Feb):194.

Levine, R. 1961. *Ethics and Regulation of Clinical Research.* Urban and Schwarzenberg, Baltimore, MD.

Marx, L. 1964. *The Machine in the Garden.* Oxford, NY. p. 364–365.

National Institutes of Health (NIH). 1980. *Issues in Research with Human Subjects.* U.S. Government Printing Office, NIH publication No. 80–1858. Washington, DC.

National Institutes of Health (NIH). 1977. *General Considerations for the Clinical Use of Drugs.* U.S. Government Printing Office, NIH publication No. 77–3040, Washington, DC.

Potter, E. J. 1991. A historical view: diethlstybesteral use during pregnancy: a 30–year historical perspective. *Pediatric Pathology.* 11(5):781–789.

President's Commission on Bioethics. 1978. Hearing before the Subcommittee on Health and the Environment of the Committee on Interstate and Foreign Commerce, House of Representatives, 95th Congress, 2nd Session, on HR 13662. Aug. 4, 1978. U.S. Government Printing Office, Washington, DC.

Telling Public Stories About Risk

Sharon Dunwoody
Evjue–Bascom Professor of Journalism and Mass Communication;
Head, Center for Environmental Communications and Education Studies,
University of Wisconsin, Madison
(pictured on right)

Communicating risk to the public looms as a confusing, perhaps even wildly unpredictable, process to many scientists and policymakers. People confronted with risky situations often seem to respond irrationally. Their reactions sometimes suggest that they are evaluating information in superficial and hasty ways. It is hard to see patterns in the judgments they make about whose information is credible and whose is not. Incredibly enough, they even seem to believe what they read in newspaper stories.

Others (see Hoban and Kendall, page 73) focused on how we, as individuals, perceive risks "out there." This presentation will focus on some work by communication scholars that explores how we all use information to make judgments about risk. More specifically, I want to talk about how, given a risky situation, individuals choose information channels in order to learn about the risk, as well as to decide how worried to be about that risk.

Why care about information channels? Let me respond in two ways. One is that the old–fashioned view of the risk communication process—a simple stimulus–response scenario in which the expert spouts information and the recipient ingests and then acts in ways consonant with that information—rarely seems to work, and ignoring channel preferences may be one of the many reasons why. A second reason for caring about channels is that the findings of communication research suggests their role is counter–intuitive. Put another way, we all seem to routinely misjudge the effects of channel use on people's risk judgments. If part of the goal of this report is to engage the agricultural biotechnology community in fruitful public discussion of risk, then reconfiguring our understanding of channel use is important.

...reconfiguring our understanding of channel use is important.

FIRST, AN EXAMPLE....

Before discussing some of the things we have been learning about how folks use information channels to inform their risk judgments, let me begin with an example. A very personal one, I might add.

Some years ago, Steve, my partner, and I sat down to watch a *NOVA* program about asbestos. Midway through, Steve wondered aloud if the stuff wrapped around the steam pipes in the basement of our old refurbished farmhouse might not be moldering asbestos. It was. We now had to decide what to do about it.

We began an intensive search for information. We called state agencies, the Environmental Protection Agency (EPA) and various information offices at our university. We rummaged through libraries. We ended up with a large pile of information about asbestos but, we felt, no information specific enough for our needs. So we next embarked on a search for individuals who could investigate our problem in person. Two engineers ultimately found their way into our basement. One resurfaced with the soothing message that we would be quite safe if we left the asbestos as is; the other hastened back up the basement stairs and warned us to stay out of the basement until the asbestos had been removed.

In desperation, we finally asked ourselves: Would we ever return to the basement if we left the asbestos in place? The answer was no. A few weeks later we hired the best professional we could find to remove the stuff.

Although I did not realize it at the time, this saga nicely illustrates some of the more important channel factors that scholars have discovered in recent years. Three will be discussed here: 1. the notion of channel utility; 2. the argument that individuals use different channels to inform different dimensions of risk judgment; and 3. the argument that individuals evaluate information in some channels as more relevant to themselves than information in other channels. Be aware that I make a distinction between "channel" and "source." A channel is a mode of transmission—*The New York Times, 60 Minutes*, an interpersonal interaction. Sources, on the other hand, are information providers embedded in channels. A single channel can offer many sources, or just one.

CHANNEL UTILITY

Our world is awash in information channels. The mass media are obvious ones, and they often get fingered as the sole, or at least the primary, channels used by the public to learn about risks. But recall how you came to terms with a recent salient risk and you will realize that, in an information–rich society such as this one, we have many channels at our disposal. In the asbestos example above, Steve and I utilized television, newspapers, various printed brochures and pamphlets, and human beings.

But we typically do not access these channels in equal dollops. Stanford researcher Steve Chaffee (1986) argues that our use of any particular channel depends on two things: the cost of getting to that channel and a judgment of the likely relevance of information that we may find there. The joint outcome of those two factors determines something called "channel utility."

Some channels are too costly to use, even when we judge the information they contain to be of high quality. For example, many people will cite a physician as their preferred channel for information about health risks such as AIDS (Freimuth et al., 1987), but few individuals will actually discuss those risks with physicians. The physician ranks high in terms of likely information relevance, but she is also costly to access; most individuals in our culture either cannot afford to visit a doctor just to talk about health issues or are reluctant to ask time of such a high–status person.

Conversely, other channels may be easy to access but may be judged inappropriate for certain types of information. For example, some of the most accessible publications in the country are sitting in supermarket checkout lanes. Yet, many individuals would regard *The National Enquirer* as a poor source of information about biotechnology risks.

For most of us, juggling cost and relevance leads to channel tradeoffs.

For most of us, juggling cost and relevance leads to channel tradeoffs. In fact, more often than not, we may settle for a particular channel not because we prefer it, but because it is available. Many of our preferred channels may be too costly to access and we, thus, "make do" with a variety of channels—such as the mass media—whose information we may regard as reasonable but not necessarily on–point. For example, although we may prefer to learn about AIDS from our physicians, we do not. Instead, we rely heavily on the mass media for such health information. That reliance stems not from a judgment that newspapers are better sources of health information than physicians—to the contrary, when asked, individuals are quick to note otherwise (Reagan and

Collins, 1987)—but from the fact that newspapers are far easier to access and are regarded as being generally informative.

RISK JUDGMENT AND CHANNEL CHOICES

The days when scientists went looking for a tight fit between level of risk and behavioral response are long gone. Risk perception researchers have made it abundantly clear over the years that we all use multiple factors to evaluate a risk and that estimates of likelihood of coming to harm—while taken into account when available—are only part of the picture and must compete with other factors, such as our familiarity with the risk, our sense of control over it, and its catastrophic potential (Slovic, 1987).

If one acknowledges that decisions about risky situations are grounded in multiple dimensions rather than just one, then it is a short conceptual hop to the idea that individuals may prefer different channels to inform different dimensions of risk judgment.

And that is just what risk communication researchers are finding. Recent work suggests that individuals differentiate between channels that are appropriate for *learning about* a risk and those best used to decide *how worried to be* about the risk. Specifically, individuals seem willing to rely on the mass media and other "impersonal" channels in order to gather information about a particular risk. But they seem unwilling to rely on those same channels to reach decisions about how worried to be. Instead, they prefer interpersonal channels to inform this "worry dimension."

...individuals differentiate between channels that are appropriate for learning about a risk and those best used to decide how worried to be about the risk.

For example, in a study of young adults' perceptions of the risk of contracting the HIV virus, Dunwoody and Neuwirth (1991) found that use of the mass media predicted to participants' understanding of level of risk (a cognitive, knowledge variable) but not to their level of worry about contracting the virus (an affective variable). Rather, the best channel predictor of level of worry was interpersonal.

While this flies in the face of a pervasive cultural assumption that the mass media can scare us to death by "sensationalizing" information, it is quite consistent with a growing body of mass communication research that finds media messages far more closely linked to cognitions than to affect. That is, the media seem to operate principally as sources of information in our world, not as persuasive forces.

That distinction played itself out in the little asbestos saga above. Steve and I gathered lots of mass-produced information and learned a great deal about asbestos. But we were reluctant to use that information to construct a sense of how worried we should be about the stuff in our basement. Instead, we wanted to talk to human beings. We sought out interpersonal channels to help us with that dimension of risk judgment.

REFERENTIAL LEVEL AND CHANNEL CHOICES

Risk perception researchers have demonstrated that, when confronted with a hazard, we systematically differentiate between the level of risk it poses to others and to ourselves. Specifically, we underestimate our personal level of risk compared to that of others (Weinstein, 1989).

Again, it is a short conceptual hop to the idea that, if individuals distinguish between self and others when judging level of risk, they may utilize different information channels to inform those two different judgments.

And again, mass communication research bears this. Specifically, individuals seem to interpret mass media information as telling them about the risks to people "out there" but resist seeing those same messages as telling them anything about their personal level of risk. Instead, once again interpersonal channels are the preferred source of personal risk information.

Two studies offer good illustrations of this phenomenon. Tyler and Cook (1984), in a series of experiments examining the ways in which information influenced individuals' judgments of the risk of being victimized, found that: 1. personal– and societal–level judgments were quite independent of one another; and 2. mass media crime stories influenced those societal–level judgments but not the personal ones. In other words, reading newspaper stories about crime leads you to think that folks around you (i.e., in your community, your state, your country) have a greater likelihood of being victimized, but the stories will not influence your assessment of your own personal level of risk. The world around you may look scarier, but you see your neighborhood as immune to that trend.

...interpersonal channels are the preferred source of personal risk information.

Similarly, Culbertson and Stempel (1985), in a survey of Ohio residents, found the self–other distinction: Seventy–five percent of the respondents criticized the availability of health care in the United States while only 5 percent viewed their own health care environment negatively. Further, evaluations of media coverage of health were correlated more with respondents' assessments of the quality of health care available to Americans than with their assessments of their personal care.

In sum, mass–mediated information influences our social–level perceptions but not our individual–level ones. This differential impact has come to be known as the "impersonal impact hypothesis," as it suggests that, in the words of Tyler and Cook (1984), "the modality of indirect experience, which is most effectively controlled by society and which reaches the largest audiences—the mass media—is the least effective in influencing personal concerns" (p. 694).

This referential distinction emerged starkly during the asbestos saga. While Steve and I gathered reams of written information about asbestos, we resisted seeing the information as telling us something about our personal level of risk. We clung to the assumption that our situation was somehow unique. The more general written documents could not be sensitive, we felt, to the amount, age and condition of the asbestos in our basement. We finally resorted to bringing in human beings to stare at our asbestos and offer recommendations for dealing with it.

Again, the argument that mass media channels are ineffective at the personal level flies in the face of many individuals' assumptions about our use of information channels. We all believe we have witnessed the impact of the media on personal perceptions, be it the specter of thousands of residents who fled their homes in 1979 in reaction to the news that the damaged Three Mile Island reactor might harbor a potentially catastrophic hydrogen bubble or the hundreds of phone calls to cancer or AIDS hotlines after the inevitable public disclosure of a well–known personality with the disease.

...individuals use newspapers, radio and TV as social antennae to alert them to situtations or issues "out there" that may need attention.

Indeed, a good bit of empirical evidence suggests that the mass media do serve an alerting function, that individuals use newspapers, radio and TV as social antennae to alert them to situations or issues "out there" that may need attention. But once the issue has become salient to an individual, that "agenda–setting" function is only the beginning of an elaborate process of information–gathering, one in which the mass media are only part of a panoply of channels, each being used for very specific but very different purposes.

By way of example, I turn again to the asbestos saga. It was television—specifically a *NOVA* program—that alerted Steve and I to the problem of asbestos. But when we began looking in earnest for detailed information about the risk, we gave little thought to seeking information in media channels. Media stories are time–based, intermittent, ephemeral. They lack detail. Their presence coincides with news "out there," not with the personal situations of their readers. So for us, the mass media served its classic alerting function and then vanished as a relevant channel during the rest of our search.

THE LIMITS OF MASS MEDIA CHANNELS

This inability of media channels to inform individuals' personal levels of risk is disconcerting to some policymakers for another reason: They engage in multimillion–dollar information campaigns to convince us to change a number of life–threatening habits, from smoking to having unprotected sex, and those campaigns traditionally rely heavily on the mass media to carry their messages. Our resistance to seeing mediated messages as relevant to ourselves is costly to campaign designers—so costly that it has sparked a good deal of discussion about why the public makes this channel distinction and what can be done about it. Researchers have proffered a few educated guesses.

One argument is that our insistence on interpreting mediated channels as informing only our understanding of society is a learned response applied to all media messages. That is, we have all have grown up amidst the mass media and, over time, have learned that media stories are always about folks "out there," never about us. We have learned, in other words, to interpret media messages as telling us about others, about society.

If this argument is correct, then one may be able to counteract that pattern either by recasting media messages in ways that signal to the reader that "this story is about you" or by training audience members to interpret existing messages differently. In the former camp, strategies might include beginning stories with story narratives featuring individuals like the typical reader, or using the second–person "you" throughout the story. A focus on the latter strategy must begin with a better understanding of how people interpret mediated messages and then would require a kind of resocialization process. Efforts to change audience perceptions through restructuring media accounts have not been fruitful to date (see Dunwoody et al., 1992). Although scholars have not yet explored the notion of "retraining" media message users, researchers have demonstrated the value of educational efforts in such areas as promoting better individual use of mathematical and statistical concepts (see Nisbett et al., 1983).

...recasting media messages in ways that signal to the reader that "this story is about you"...

Another argument is that individuals do indeed use media channels to inform personal risk levels but that, since most of the risks reflected in media accounts are those whose likelihood of occurrence is low, no change in personal risk levels is necessary. Tyler and Cook (1984) maintain that, under such conditions, "the rational and discerning response of most members of the public probably should be to refrain from changing their estimation of their own risk while acknowledging that the problem may be greater to society in general than they had thought" (p. 206). This hypothesis has not been tested, but the self–other distinction seems to hold across a range of risks and likelihood levels.

Finally, yet a third group argues that we are so resistant to seeing ourselves as being at risk that nothing can dissuade us from interpreting our level of risk as lower than that of the folks around us. If true, no amount of fiddling with media stories about hazards will convince audiences that those stories have something to say about their personal likelihood of coming to harm. Partial support for this position comes from Gunther and Mundy (1993), who found in one study that media stories recounting disadvantageous consequences generated the self–other referential distinction while stories that posed potential benefits did not. Respondents immediately interpreted the positive stories as relevant to themselves, in other words, but resisted seeing the negative stories in the same way. This suggests that at least part of the problem lies with our reactions to the message rather than to the channel.

DISCUSSION

Learning how people use information to inform their risk judgments is difficult. A literal blizzard of factors about the individuals themselves can influence those judgments, everything from a person's available store of knowledge about the risk to personality factors that make some people more likely to take risks than others. Attributes of the messages themselves introduce another welter of factors, from the clarity of the words and phrases to the vividness of the text.

But only a very few, socially advantaged souls have such extensive channel access.

In this brief presentation I have tried to illuminate one element within that panoply of message factors: the influence of channel. The bottom line here is that channel makes a difference. Given channel choice, individuals will use different channels to help them make decisions about different dimensions of a risk. For example, a magazine article about radon may contribute to their understanding of the damage that radon can do to the human body. But when it comes to deciding whether or not to install a radon detector—that is, whether individuals should be worried enough about the risk to engage in some level of expense to determine the level of hazard to themselves—they will opt for a channel that they feel can take their personal situation into account. Almost without exception, that channel is interpersonal.

What does this mean for risk communicators? First, I think it requires us to be clear about our communications goals and to select channels that fit with those goals. You may employ very different channels to *explain* a risk—actual or potential—than you will use to try to *persuade* audiences that the risk should or should not worry them.

Second, it forces us to ponder the inequitable nature of channel access. In an ideal world, individuals could select among a myriad of channels to meet their informational and decision needs. They could locate both popular and technical documents in publications or in electronic databases; they could talk to experts.

But only a very few, socially advantaged souls have such extensive channel access. Many Americans live their lives in something of an informational straightjacket; economics and the social context within which they live have severely restricted their channel choices. They do not subscribe to a newspaper. They may buy only the occasional magazine off the newsstand. They have little experience with libraries, even less with searching for information by computer. They have no idea how to get to human experts. They have never made a phone call to a governmental agency in search of an answer to a question.

For these individuals, the cost of using even generally available channels to inform their risk judgments may be high indeed, so high that it serves as an effective barrier to informed decision–making.

Finally, research on channel use raises the specter of an active audience. Turn–of–the–century communication researchers viewed the audience as a passive mass that absorbed and reacted to messages in predictable ways. Studies since World War II have turned that image around, suggesting instead that information users play an important role in selecting and processing messages. That filtering process can make or break a communication attempt, and it means that risk communication managers must be sensitive to the information recipient as a major player in the communication process.

REFERENCES

Chaffee, S.H. 1986. Mass media and interpersonal channels: competitive, convergent, or complementary? In *InterMedia* (3rd ed.). G. Gumpert and R. Cathcart, eds. Oxford University Press, New York, NY. p.62–80.

Culbertson, H.M. and G.H. Stempel III. 1985. "Media malaise": Explaining personal optimism and societal pessimism about health care. *Journal of Comm.* 35:180–190.

Dunwoody, S. and K. Neuwirth. 1991. Coming to terms with the impact of communication on scientific and technological risk judgments. In *Risky Business.* L. Wilkins and P. Patterson, eds. Greenwood Press, Westport, CT. p. 11–30.

Dunwoody, S., K. Neuwirth, R.J. Griffin and M. Long. 1992. The impact of risk message content and construction on comments about risks embedded in "letters to friends." *Journal Lang. and Soc. Psych.* 11:9–33.

Freimuth, V.S., T. Edgar and S.L. Hammond. 1987. College students' awareness and interpretation of the AIDS risk. *Sci. Tech. & Hum. Values.* 12:37–40.

Gunther A.C. and P. Mundy. 1993. Biased optimism and the third–person effect. *Journalism Quarterly.* 70:2–11.

Nisbett, R.E., D.H. Krantz, C. Jepson and Z. Kunda. 1983. The use of statitical heuristics in everyday inductive reasoning. *Psych. Rev.* 90:339–363.

Reagan, J. and J. Collins. 1987. Sources for health care information in two small communities. *Journalism Quarterly.* 64:560–563,676.

Slovic, P. 1987. Perception of risk. *Science.* 36:280–285.

Tyler, T.R. and F.L. Cook. 1984. The mass media and judgments of risk: Distinguishing impact on personal and societal level judgments. *Journal Personality and Soc. Psych.* 47:693–708.

Weinstein, N. 1989. Optimistic biases about personal risks. *Science.* 246:1232–1233.

Communicating With the Public About Risk

Jerry E. Bishop
Deputy News Editor
(science)
Wall Street Journal

As many biotechnologists, particularly those involved in agriculture, are beginning to discover, it is not easy to communicate to the public the concepts of risk, at least not risk as understood by the scientific community. Many people find themselves baffled and chagrined by a weather forecast that says there is a 30 percent chance of rain when all they want to know is whether to take an umbrella to work.

Those who toil in the mass media are quite conscious of the difficulties of communicating risk concepts to the lay public. Hardly a week goes by when editors and reporters are not forced to decide whether to publicize some alleged threat to the public health. The threat may range from the risk of too little calcium or too much iron in the diet to the risk of using a cellular telephone or driving a pickup truck with the fuel tanks mounted outside the truck frame. Inevitably, one interest group will accuse the media of needlessly scaring the public while another interest group will charge that the media are failing to alert the public to a deadly danger.

The criticisms of the media from the scientific community, however, are of a different nature, are considerably more reasoned and, consequently, are more closely attended than those from special interest groups. One such critic is Daniel E. Koshland Jr., the esteemed editor of *Science.* In an editorial in late 1991, Koshland had this to say about the mass media:

> There are many examples these days of improper conduct, of which the recent coverage of the chemical Alar, used to slow the ripening of apples, is a dramatic example. In that case, a clearly dubious report about possible carcinogenicity by a special interest group was hyped by a news organization without the most simple checks on its reliability or documentation. This caused panic among consumers and losses of millions of dollars by apple growers. Confronted with the inadequacy of the data, a spokesman for the public interest group recently suggested that it was excusable because people are eating more apples than ever before. That is like an embezzler justifying embezzlement by saying the banking industry continues to survive. Worse, the public's disdain for repeated scares indicates that an individual publication's (or broadcast group's) willingness to cry 'wolf' uncritically may be destroying the press's own credibility and its ability to provide legitimacy to responsible environmentalists...the press has been too willing to publicize Jeremy Rifkin's cries of alarm, which so far have been consistently wrong.

Koshland goes on to argue that the press should adopt a policy of revealing the sources of data that are claimed to be "scientific" and should distinguish between a report in a peer–reviewed scientific journal and the claims made at a "dataless press conference" or in "a public relations document." He urges that "press conferences without peer–reviewed data should be greeted with heavy skepticism."

Most scientists, particularly those in biotechnology who have been in the glare of Jeremy Rifkin's pronouncements, probably agree with Koshland's recommendations on how the press should perform. But Koshland begs the more basic question of why a reputable organization like CBS Television and a highly regarded program like *60 Minutes* would deign in the first place to scare the wits out of the apple–eating American public by publicizing a report that lacked peer–reviewed scientific data.

To understand why television, newspapers, news magazines and other media would publicize such an unsupported allegation of a health danger it might be useful to look at a few examples. None, in this case, deal directly with agricultural biotechnology but they offer an insight that might be useful to those who might have to deal with the media and the public about issues of safety and risk of genetically altered crops and irradiated foods.

An interesting piece of history appeared a few years ago in the journal *Preventive Medicine.* Andrew McClary, a science historian at Michigan State University, had dug into newspapers and consumer magazines published in the early part of this century to see how they dealt with the problem of the housefly (McClary, 1982). The germ theory of disease was fairly new at the time and early research had discovered pathogens were harbored in the gut of the common housefly. It was well known, of course, that houseflies feed by regurgitation and that they are commonly seen flitting from outhouses and garbage cans to kitchens and dining rooms. These observations had led to the seemingly logical conclusion that the ordinary housefly could spread disease.

The evidence that the housefly was a vector of human disease was purely circumstantial. There had been no documented cases of human illness being directly transmitted via the housefly and infectious disease experts of the day said it seemed unlikely that the housefly was an important health hazard.

Nevertheless, the mass media fell in whole–heartedly with local campaigns to warn the public of the dangers of the housefly. In addition to many articles about proper sanitation, many newspapers championed "swat the fly" campaigns, McClary found. In Washington, D.C., 5,000 children brought in an estimated 7 million dead flies during a two–week campaign in which the *Washington Evening Star* offered prizes up to $25 for the most flies killed. The champion, 13–year–old Layton Burdette, brought in 343,800 dead flies, having paid a company of 25 boys to kill and collect flies for him.

Inevitably, one interest group will accuse the media of needlessly scaring the public while another interest group will charge that the media are failing to alert the public to a deadly danger.

It was the consumer magazines of the day, however, that went after the housefly with a vengence, McClary found. In 1911, the *Independent* described the fly as "a monstrous being with more eyes than Argus, wings like a monoplane, six long, hairy legs and a mouth armed with horrid mandibles, sucking blood and dripping poison."

McClure's charged, in 1909, that the fly would "slaughter the little babies through the summer." A year later, *Ladies Home Journal* asked its readers: "What will you do? Shall he continue in his death–dealing path or will you rise and 'swat' him?" In 1913, *Good Housekeeping* declared that "women are the mothers of babies and the makers of homes, and the fly is an enemy of both."

These exaggerated assertions undoubtedly caused the public an immeasurable amount of anxiety and led to the expenditure of an enormous amount of time and money to eliminate the housefly—an effort that continues to this day. But the media went unchallenged. Rarely, if ever, did infectious disease experts and public health officials step forward and accuse the media of scaring the public on the basis of inadequate scientific evidence. Unlike the Alar controversy, the mass media's "play" of the housefly menace did not damage any particular interest. The housefly had no defenders; there was no insect rights group trying to protect the innocent housefly. Fly control even spawned a not–so–small industry in window screens, fly paper, fly swatters and, years later, insecticides.

So it would seem that one aspect of communicating risk to the public is whether the communication is likely to affect some particular economic interest. Koshland's editorial condemns "a clearly dubious report...by a special interest group"—presumably the Natural Resources Defense Council—because it "caused panic among consumers and losses of millions of dollars by apple growers" but he fails to classify the apple growers as a special interest group. One can only wonder whether the Alar story would have stirred such condemnation if it had not caused a precipitous—but temporary—decline in consumption of apples and apple juice.

Be that as it may, McClary raises a question regarding the housefly articles that remains pertinent to this day:

> Should one condemn these articles as failing to meet desired standards of popular science writing? Was it better to gain reader interest through mild sensationalism, or risk its loss through the use of unemotional, objective prose?

This is not a trivial question that applies only to the media in the early part of this century. It is particularly pertinent today when a potential news story deals with "risk."

Every reporter and editor knows that if a story fails to interest the reader, the reader simply turns the page and looks for some other story that does interest him or her. No reporter is going to waste his or her time and effort writing a story that no one will read. And any newspaper or magazine or television news program that consistently publishes articles that fail to interest readers will quickly discover its readers going elsewhere for information.

At the same time, readers do not like to be misled; they resent it when they invest their time in reading a story that turns out to be far less interesting and informative than promised by the headline and "lead" of the story. One can imagine the chagrin of readers who, a few years ago, spent a dollar on a supermarket tabloid with a headline "Man Shot Eight Times and Lives" only to discover the story is about a body found with nine bullet holes.

In news stories dealing with risk, the reporter and then the editor have to decide how to arouse a reader interest enough to make them pause and read the story and yet not mislead the reader. I was reminded recently how difficult it is to tread this line by an incident involving one of my own stories (Bishop, 1993), an incident which, I fervently hope, continues to be rare on *The Wall Street Journal.*

A medical journal had published an article describing a long–term follow–up of women who had had chest irradiation after surgery for breast cancer. The analysis indicated that 10 years after irradiation there was a two–fold increase in risk of lung cancer. Any story about breast cancer inherently stirs reader interest. Many readers have either had breast cancer or have family members or friends who have had breast cancer. And this story would carry particularly strong reader interest because in recent years many women diagnosed in the early stages of breast cancer have opted for a so–called "lumpectomy"—plus radiation—in hopes of preserving the breast.

...how to arouse a reader interest enough to make them pause and read the story, and yet not mislead the reader.

But the report lacked certain information. It described only relative risk. The absolute risk of a woman developing lung cancer after breast irradiation was not calculated. Moreover, the effect of cigarette smoking on the relative risk of lung cancer was not taken into account. A call to the authors revealed that their main interest was not in the safety or long–term effects of breast cancer therapy, but rather in gathering evidence on the induction of cancer by ionizing radiation. The lead author had had several calls from the press and was becoming a bit overwrought that news stories might unduly influence therapeutic decisions for women with breast cancer.

We assumed that many of our readers would hear of this study on the evening news or read it in their morning paper. We also felt that if the story were not presented in the proper context, women readers who have had or might have breast cancer would be unduly alarmed about the findings. Therefore, it was decided that we would present the new findings to our readers in a context that was informative but not alarming. To this end the first sentence of the story, the "lead," read:

> A Columbia University scientist cautioned that his finding that radiation treatments for breast cancer increase the risk of lung cancer were mainly of scientific interest.
>
> "I don't think women who are being treated for breast cancer or who had radiation treatments for breast cancer in the past should be overly concerned" about the new finding, said Alfred I. Neugut, a cancer specialist at the College of Physicians and Surgeons, Columbia University's medical school in New York.

Later in the evening, an editor on the copy desk decided that the lead sounded as though the finding of higher lung cancer risk was old news; it read as though the reader already knew about the finding. He changed the lead as follows:

> A new study says radiation treatments for breast cancer increase the risk of lung cancer, but the Columbia University scientist who headed the research team said the finding was mainly of scientific interest and should not alarm women who receive the treatment.

This version subtly puts more emphasis on the increased risk of lung cancer. The change, however, is not too far from the original and still cautions the reader that the study should not affect decisions on breast cancer therapy. However, a headline has to be written that sums up the story in few succinct words. And the headline usually is a condensed version of the lead of the story. Thus, the next morning the story appeared with this headline:

STUDY LINKS BREAST CANCER TREATMENT TO HIGHER RISK OF THE DISEASE IN LUNGS

The headline, which gives readers a certain "mindset" before they read the story, states exactly what we originally tried to avoid. But the episode shows how difficult it is sometimes to grab a reader's interest without resorting to the melodramatic exaggerations that the anti–housefly writers used three–quarters of a century ago.

News of a risk attracts more interest than news of no risk.

The editor's reaction to the original lead also underscores another aspect of reporting risk—oriented stories. News of a risk attracts more interest than news of no risk. The original lead did not appeal to the editor, in part, because it indicated that readers need not concern themselves about the finding. The change put more emphasis on the discovery of a previously unknown risk and the resulting headline stated it rather baldly.

This bias against "no risk" stories was described in late 1991 in a report appearing in the *Journal of the American Medical Association*, "Bias Against Negative Studies in Newspaper Reports of Medical Research" by Koren and Klein of the Hospital for Sick Children in Toronto (Koren and Klein, 1991). Earlier in the year a single issue had carried two articles on risk of cancer among populations exposed to radiation. One study, the "positive" finding, showed that atomic energy workers at the Oak Ridge National Laboratory had 63 percent higher than normal risk of developing leukemia. The other study, the "negative" study, failed to find any increased risk of cancer among people living near nuclear power plants.

Koren and Klein found 19 daily newspapers that had carried stories about the articles but only 10 of these reported the results of both studies and, of those 10, most emphasized the "positive" finding of a leukemia risk among Oak Ridge employees and gave minor consideration to the "negative" finding of no risk. Nine of the newspapers reported only the "positive" finding.

Readers...want to know what such criticisms are and when they are expressed.

Koren and Klein conclude: "Responsible journalists should acknowledge the importance of providing balanced information to the public when covering controversial health issues and should give equal attention to positive and negative studies."

Koren and Klein's finding of bias against negative studies is not surprising. What Koren and Klein actually uncovered was not so much a bias on the part of the newspapers as a bias on the part of newspaper readers. News of health risks must compete for reader attention each day against news of other happenings in the world—including wars, murders, economic declines, election campaigns, Congressional votes on health care issues, stock market gyrations and the agonies and accomplishments of sports teams. It seems obvious that a scientific study that reveals a potential threat to one's health will compete well for the reader's attention. On the other hand, a report that tells the readers they need not worry about getting cancer from living near a nuclear plant will be of little interest to people who do not live near a nuclear power plant, which probably includes most of the readership of the newspapers surveyed by Koren and Klein.

It is somewhat naive to assume that people will read whatever the press decides to print. The fact is that, unlike a teacher or even a preacher, the audience served by most publications is not a captive audience that is required to sit there and read every word impressed on the page. It is a capricious audience, a fickle audience that picks and chooses what it wants to read.

First and foremost readers will want to know how the new technology will affect their lives...they will immediately want to know if it is in any way harmful or dangerous.

This is the overriding criterion used by editors in deciding what news will be printed about risk or any other subject.

This criterion applies just as stringently to news about new technologies such as the genetically engineered economic animals and crops which the NABC audience deals with. First and foremost readers will want to know how the new technology will affect their lives. Thus, they will have more interest in genetically engineered tomatoes that will appear on their supermarket shelf than a strain of corn genetically engineered for drought resistance. Initially, they will be most interested—and the press will be most likely to report—how the new technology will benefit them, the readers.

But once the readers learn that they soon will encounter the products of a new technology and that its inventors are promising benefits, they will immediately want to know if it is in any way harmful or dangerous.

The press is acutely aware of these desires of readers and will publish the information they want as it becomes available. The first stories will describe the new technology, its potential benefits and probably include the assurance of its developers that it is safe. But the reporters and editors will be on the alert for any indications that the new technology might carry known or unknown risks. The risks, incidentally, could be financial as well as health–related since the readership includes those who might want to invest in the new technology.

This, for instance, is the reason the press gives publicity to pronouncements and actions by critics like Jeremy Rifkin and Ralph Nader, and prints stories about chefs of famous restaurants agreeing not to serve genetically engineered foods. Readers, including many attending NABC 5, want to know what such criticisms are and when they are expressed. It is as important, or perhaps more important, for the scientists and venture capitalists involved in developing genetically engineered foods to know what Jeremy Rifkin and the chefs are doing as it is for the general public.

To use any other criteria for deciding what should be printed or not printed in a newspaper or any other medium of mass communications would be lethal for the newspaper. To paraphrase one of my editors of several years ago, the newspaper editor who decides to print what he thinks people should read instead of what they want to read will soon find he has no newspaper to be editor of.

REFERENCES

Bishop, J. E. 1993. Study Links Breast Cancer Treatment to Higher Risk of the Disease in Lungs. *Wall Street Journal.* May 14, p.B6.

Koren, G. and N. Klein. 1991. Bias Against Negative Studies in Newspaper Reports of Medical Research. *J. Amer. Med. Assoc.* 266(Oct 2):1824–1826.

Koshland, D.E. Jr. 1991. Credibility in Science and the Press. *Science.* 254(Nov 1):629.

McClary, A. 1982. "Swat the Fly": Popular Magazines and the Anti–Fly Campaign. *Preventive Medicine.* 11(May):373–378.

PART IV

Roundtable

Martin (moderator): Good evening ladies and gentlemen. I am delighted to welcome you to NABC 5. The title of this conference is "Agricultural Biotechnology: A Public Conversation About Risk." Instead of having an after dinner speaker this evening, we will have a roundtable discussion regarding public concerns about biotechnology and its potential risks and benefits. I will follow a talk show format and interact with a panel that I will introduce to you in just a moment. By the time we conclude later this evening I hope you will be saying "I am glad that I came. I am glad to be a part of this conference. I am looking forward to hearing the plenary session speakers. I am especially anxious to enter into a dialogue in each of the four workshops on Thursday and Friday." This evening's panel will offer an opportunity to set the stage, to excite you, and to increase your interest in the conference theme: "Agricultural Biotechnology: A Public Conversation About Risk."

Let me briefly describe how we are going to proceed. I have a basic script that I intend to follow. I have shared with each of the panel members the general theme, but they don't know the questions that I will ask, so the discussion will be extemporaneous.

Let me introduce our panel members. **Lilly Russow** comes from the Department of Philosophy at Purdue. Seated next to her is **Rosetta Newsome** from Chicago, where she is with the Institute of Food Technologists and is Director of Scientific Affairs. Seated next to her, is **Rebecca Goldburg** from the Environmental Defense Fund in New York. **Bill Greenlee** is the Head of the Department of Pharmacology and Toxicology in the School of Pharmacy at Purdue. **Ann Sorensen** comes from the American Farmland Trust where she is the Director of the Center for Agriculture in the Environment. **Ted McKinney** is from Indianapolis. He works for DowElanco, a relatively new joint venture between the Elanco Division of Eli Lilly and the Dow Chemical Company. **Karen Bolluyt** is from Iowa State University, where she is the Director of the Agricultural Information Service. **David Judson** is a journalist with the Gannett News Service, based in the Washington, DC area.

As you can see, we have people with very interesting and diverse backgrounds. You will have a chance to work closely with them on Thursday and Friday. They are going to serve as cochairpersons for the four workshops. You also will hear from each of them in the wrap–up Session on Friday afternoon.

With this introduction, let us begin the evening program. Everyone travelled to this conference. Of course, some of you came from further away than others. Some only had to travel a short distance in West Lafayette, while others came from various places across the United States. No one chose Amtrak, although we do have service in Lafayette from Chicago to Indianapolis. How many came by commercial plane? About half the group. How many of you drove? Another half the group. Rebecca Goldburg, why did you decide to fly?

Goldburg (Environmental Defense Fund): Mainly because it takes far less time than any other mode of transportation. Also, it's relatively safe.

Martin: Good, she gave us two reasons: relatively safe and efficient in terms of time. Let me ask Ted McKinney the same question. Why did you decide to drive?

McKinney (DowElanco): The same reasons. It is far more efficient to drive the one hour from Indianapolis, and it's fairly safe.

Martin: David Judson, you came from Washington, DC. Why did you choose to fly?

Judson (Gannett News Service): It's a logical, efficient, quick way to get here.

Martin: Karen Bolluyt, did you drive over from Ames or fly?

Bolluyt (Agricultural Communications, Iowa State University): I flew.

Martin: You flew. There are some people here from Ames who drove. It took them about eight hours to drive. You chose to fly. Why?

Bolluyt: Because it was quick.

Martin: Rosetta Newsome, did you drive from Chicago?

Newsome (Institute of Food Technologists): Yes, I did, but had I realized there was Amtrak service, I might have considered that.

Martin: One of the choices that these individuals made in coming to this conference was how to get here. Some of you considered economics; some considered the time and convenience factors. And some considered risk. Let me give you some statistics about travel risk. In the United States, there were 515 fatalities in 1989 for those folks who traveled by Amtrak. If you'd known that, Rosetta, do you think that you would have come by Amtrak from Chicago?

Newsome: No, I would have considered the ease of getting here.

Martin: Also in 1989 there were 278 fatalities on U.S. commercial airplanes. Someone commented that travel by air is relatively safe. Does anyone want to guess about how many automobile fatalities there were in 1989?

Goldburg: 50,000.

Sorensen (American Farmland Trust): Much higher...

Martin: Like 100,000 maybe.

Russow (Philosophy Department, Purdue University): At least.

Martin: What do you think, Dave?

Judson: I'd say 75,000.

Martin: There were 46,000 passenger car fatalities in 1989. Ann, I understand that you drove down. I recall that in visiting with you on the telephone you said you prefer not to fly.

Sorensen: I am a nervous flyer; I am well aware that flying is safer than driving. It's an irrational fear of flying, but I drive when I can.

Martin: So here we have a scientist that knows the facts about the risks of travel but tells us that sometimes emotion plays a factor. We're going to talk about this during this conference. Let me move on now to another issue. Many Americans are concerned about food safety, particularly pesticide residues. Let us talk about food safety. I did some shopping. I bought a tomato. As far as I know, this one was produced with the application of pesticides. Lilly, would you eat this tomato?

Russow: I wouldn't buy it at a grocery store, but I would eat it at a restaurant.

Martin: Did it worry you that the salad you ate at the banquet tonight had tomatoes in it that might have been grown with pesticides?

Russow: Not really, I don't think it's much of a risk. I think it's much less of a risk than I took driving over here. But when I do have control over things, I don't buy things like that.

Martin: So, even though you know some of the scientific facts, you do make choices based on factors other than science.

Russow: I do.

Martin: If you had a chance to get a tomato from your garden or from a neighbor's garden that you knew was grown with minimal or no pesticides, would you prefer that tomato over one from the store that might have some pesticide residue.

Russow: Yes.

Martin: Let me turn to Rebecca Goldburg and ask the same question.

Goldburg: I think my response is quite similar to Lilly's. When I can buy food without pesticides, I try to but I certainly often buy foods that were grown with pesticides.

Martin: So you weigh in your own mind the risks and try to make a reasonable choice.

Goldburg: Yes, plus price and other factors.

Martin: Remember, David Judson is a writer for the Gannett News Service, and he often writes about science. David, let us assume that you have been assigned by your editor to write a news story about tomatoes and pesticide use. How would you tell the story? What do you think the issues should be, particularly about risks that may be associated with this food?

Judson: I first would raise the question: Is the tomato good or bad for you? I think the context is probably the most important thing, i.e., the risk of eating a tomato against the risks of driving a car, of living in a house with lead-based paint, of living in a city like Washington, DC, or whatever. There are risks all over.

Martin: Ted McKinney is trained in agriculture and works for DowElanco, a company involved in agricultural chemicals. How does your company inform the public about the relative benefits and risks of the use of pesticides in tomato production, or in agriculture in general?

McKinney: It is particularly challenging because of the public perception. I think the perception is so strong about the perceived negative aspects of pesticides relative to many other things that we all use in our daily lives that it is a huge task. We don't believe in industry that we can afford to just turn on the airways like a media campaign and move the dial over because 51 percent doesn't win. As a result, we try to build coalitions; we try to work with those who are informed such as the media. We believe that facts and science will win despite what seems to be an overwhelming task as it relates to public perception.

Martin: Let me reach into my sack again. I found another tomato! This one could be called "Flavr Savr™," one that enhances shelf life. The company that developed this type of tomato is Calgene. Many of you have been hearing about this, I am sure. Some of you may have worked on the research related to this type of technology. Now we are talking about biotechnology, or genetic engineering. This tomato is going to have a better flavor, more like that garden variety you were talking about earlier. And it will have a longer shelf life. Let me ask the same questions. How does that sound to you, Lilly?

Russow: It depends on how it was grown. I want to know where it came from and about pesticides used on it.

Martin: So information is important to you.

Russow: Yes.

Martin: Rebecca, what would you say?

Goldburg: I must admit if Food and Drug Administration approves it, I am going to try it.

Martin: Rosetta, what about the food industry? How are they going to relate to this new tomato?

Newsome: I can't speak for the food industry, but I think many consumers wish to have a tomato that tastes like the garden variety. They wish to have the tomato available year–round, they are anxious to have a good–tasting tomato available in the winter. Thus, the food industry would benefit by making a product available for purchase.

Martin: Karen, let us assume you are working on a story at Iowa State University. You want to tell a story to the consuming public through the *Des Moines Register*, a widely read newspaper in Iowa. What would you would want to say about this new tomato?

Bolluyt: My job at the College of Agriculture is to talk about what we know. It is also important that I keep in mind that what we know is only part of the story. The story should show respect for people for whom that knowledge and that information is not enough. Their value system is different than the one which is motivated primarily by what we know about the tomato in scientific terms.

Martin: I did some more shopping. I found another tomato. This one also is a genetically engineered tomato. *Bacillus thuringiensis (Bt)* protein gene has been introduced into this tomato. I don't need to explain to this audience what *Bt* is, or how it works, but I do want to talk to the panel a little bit about this tomato. I am going to start with Lilly again. Would you be interested in eating this tomato?

Russow: I am going to sound like a broken record. I want to know how it was grown, what's in it, and so on. I think that consumers are tired of the "here's a tomato, it's just like every other tomato; trust us...buy it."

Martin: So you would like some labeling information, perhaps?

Russow: Definitely.

Martin: Maybe at the grocery store in the vegetable section there should be an information sheet that the customer could pick up and take home. Rebecca, let me ask you the same question I asked Lilly.

Goldburg: Information would be important to me. I am actually in a situation where I have published a paper on the toxicology of *Bt*.

Martin: It is interesting that you mention toxicology. I have a question for Bill Greenlee. As you recall, he is a toxicologist. Are there toxicological issues that we should be concerned about? What are some of the issues we should look at to assure the public food is safe and to help them make more informed decisions.

Greenlee (Pharmacology and Toxicology Department, Purdue University): I think one of the issues that was a recurring theme in both Ted and David's comments was the potential adverse impacts on human health. The other issue is how it is grown. Were pesticides used? Clearly, pesticide use and its chemistry has evolved. We are using far fewer amounts of pesticides per acre today. It really boils down to an issue of dose. What is the exposure dose? What is the likely dose from the consumption of the tomato? Do you consume them all the time? Pesticides are likely to represent a very small residue on the tomato. We might ask how the fruit was handled in the store. There are a number of factors that need to be considered. This is a tremendous educational challenge, even among toxicologists. It is difficult to educate the scientists about what a dose response curve is. We have a tremendous challenge with the public because they see things as all or none.

Martin: This tomato grown with the *Bt* characteristic does not require insecticides to control most insects. But, a new protein has been introduced into the tomato. Rosetta Newsome, from a food industry perspective what might be important in terms of information, education, allergen concerns, or other considerations that might be important to share with the public?

Newsome: Well, we would want to share the development of the tomato. Of course, if the tomato looks like a tomato we want to be sure we are actually providing a tomato to consumers. If we are not, we need to let them know. But I think along the lines that you were speaking there are further factors that need to come into play. If this tomato is indeed a tomato, and has similar nutritional qualities to the traditional product, and does not have any known allergens induced, then we want to consider the benefits of consuming the product. We know we need to eat increased amounts of fruits and vegetables to protect our health. And this would provide benefits along that line.

Martin: Let me turn back to Ted McKinney for a minute before we move on to something else. This tomato represents developments in other crops where we may see the *Bt* characteristic to prevent insect damage. It may be in soybeans, corn, cotton or others. Your company, DowElanco, has a long tradition of producing chemicals for insect control. Now, this comes along. How does that affect a company like yours in terms of strategies or planning? Do you feel threatened by it? Are you jumping into the biotechnology business? How are you going to respond?

McKinney: If you look at our traditional product lines, you could say that it represents a threat. But if you are a well–managed company, you are looking at biotechnology as perhaps the next generation of products that would replace traditional chemicals, e.g., fungicides, herbicides, insecticides, etc. And, in fact, that is what we are doing; we are looking at this kind of thing as well. But these are not panaceas. It takes years and years even with the traditional chemistry to develop and receive government approval of new products. It takes about 10 years to register a product, 100–plus tests, and the hurdle rates are enormous. I take comfort in the science that goes into new products. Biotechnology represents new ground and, admittedly, a lot of questions must be answered from a safety and consumer standpoint.

Martin: So you suggest that regulations, although sometimes frustrating, play an important role. They provide assurance that a product is safe to the environment, to consumers, to the producers, and others involved.

McKinney: Yes, I would state very clearly that I think reasonable regulations are an absolute necessity. In fact, Becky, you said that people need to look to FDA because not everyone is going to be as informed or able to get the information at the local shopping store that guarantees how that product is produced. That is why we have regulatory agencies. So, absolutely, we support them.

Martin: OK. Who would like to have my *Bt* tomato? Ted wanted the one produced with pesticides.

Judson: I would like to have it.

Martin: I now want to turn to a less controversial topic. I would like to talk about milk! Lilly, would you comment briefly, from your perspective in the area of philosophy, if you think there are different ethical or philosophical concerns with the application of biotechnology to crops versus animals?

Russow: I think that there are two. Specifically, nobody is worried about how the tomato feels about having various biotechnology products introduced into the plant, or the seed, or whatever, but certainly there are concerns about how the cows fare in terms of their welfare with the growth hormone, bST (bovine somatotropin). And, there is a concern about whether or not it has an impact on the quality of life for cows. Another issue has come up that really doesn't have anything to do with the difference between plants and animals. That concern is with the impact on family dairy farms. I think that is very interesting. It is not just a scientific issue anymore. It has to do with societal concerns in general; what kinds of lifestyles are important and what sorts of lifestyles society wants to protect or enhance.

Martin: If I understood you right, you mentioned at least two things. One is the impact that the technology may have on the animal, itself, and whether it enhances or not the well–being of the animal. Secondly, there may be a concern about the economic, and perhaps social, impacts on farmers, of this technology. You mentioned animal agriculture in your example. What about the issue of manipulating genes, whether they are within different species of plants, within different animals, or maybe transgenic? Does this raise concerns from an ethical point of view?

Russow: I think that that is a very difficult issue, and a very muddled one because we are not very clear on the whole issue of species and species boundaries. Many people have a very strong emotional reaction once you introduce human beings. The introduction of animal genes into human beings, or human genes into animals, I think, worries the general public. There is less public concern about moving genes among plants. I think it is a concern among philosophers, because we don't know what is going on. We don't know whether species ought to be important or not. There are people who worry about playing God and messing up species, but that presupposes something about species being fixed. If you talk to biologists, they tell you the whole notion of the species is a very confused notion to begin with. It is not a very clear–cut issue at all.

Martin: Let me give you an example of something that happened to me several years ago. A scientist was working with somatotropin to enhance productivity in fish. After a speech, someone in the audience asked this scientist "Where did you get the somatotropin?" He said, "Oh, I had some human growth hormone in the laboratory and I just used that." What do you think about that? How do philosophers deal with that? Is that something that would concern them?

Russow: Personally, I think the question of where the somatotropins came from is not particularly important. On the other hand, some people are worried about the slippery slope; that is, as soon as you start harvesting human products of one sort or another, how far are you going to go? And, you get all sorts of Brave New World concerns with Frankenstein scenarios.

Martin: Like in *Jurassic Park*?

Russow: Right.

Martin: I did a little more shopping. I found this milk at a local grocery store. I did some background research on this. This milk came from a dairy farm where there are Holstein cows, so we know the breed. And it is from a herd with a 20,000 pound rolling herd average. It is a well–managed, very productive dairy herd. The dairy farmer keeps good records on the cows. I also note that this is skim milk, and it is not high in butterfat. I am going to pour a glass of milk. So, we know that this milk is homogenized, pasteurized, comes from a well–managed herd, was purchased at a local store this afternoon, and has been refrigerated. Ann Sorensen, would you drink this for me?

Sorensen: No, I am allergic to milk!

Martin: I assume you had an allergic reaction and, after some medical tests, discovered that you had a problem drinking milk.

Sorensen: Yes, I have been allergic since birth.

Martin: So here we have a case where labeling information is important. Do you buy milk in a grocery store?

Sorensen: No, I don't.

Martin: So there are times when consumer information is important. Lilly Russow, would you drink this glass of milk?

Russow: It is probably like the milk that I regularly buy in the grocery store.

Martin: So you have confidence in it based on the information on the label and the fact that there is a regulatory system to control the quality of milk. Rosetta Newsome, would you drink this glass of milk?

Newsome: I definitely would not drink it.

Martin: Why not?

Newsome: Because it has been sitting there for at least a couple of hours. Also, I don't like warm milk.

Martin: So there are many issues that we take into account when considering potential risks. Some are from a health viewpoint. Others may be questions of personal taste and preference. Would anyone else on the panel want my glass of milk?

Judson: I can drink it without any problem.

Martin: I did a little more shopping. I bought some more milk. This milk is a little different. First, let me describe what is the same. This milk also comes from a well–managed dairy farm with a 20,000 pound rolling dairy herd average per cow. Again, it is skim milk. It is cool. It has not been sitting out very long. There is only one additional piece of information that I know about this milk that I didn't about the first milk. It comes from a cow treated with bST. I have another clean glass. I will not ask Ann Sorensen to drink this milk because we already know that she has an allergic reaction to milk in general. Ted McKinney, do you have a concern with drinking this milk?

McKinney: No, I would drink it.

Martin: Why?

McKinney: Well, partly because I have been involved in research and understand the safety hurdles and tests that it takes to demonstrate bST safety. bST is found naturally in cows. I think it comes down to education. I understand the background about bST. I would feel very comfortable drinking that milk.

Martin: Karen Bolluyt, I wish to ask you a question similar to the one that I asked you before. Again, you are writing a story to be released by Iowa State University to appear in the *Des Moines Register.* How would you go about telling the story about the safety of milk, particularly milk that comes from a cow injected with bST?

Bolluyt: I hope my story would abide by two principles. One, that we are a trustworthy source of information. Two, that we respect people's ability to make up their own minds once we give them accurate, factual information.

Martin: David Judson, let me ask you a similar question. You are writing this story for Gannett. The story will be carried across the nation. You want to tell the story about bST being used in milk. How would you approach the story? Where would you seek information?

Judson: I think the points that Karen raised are fundamental. You have to ask the question "Is it safe to drink?" You have to go to sources that are credible. I think that the FDA, or similar sources, are probably regarded by most people as credible. The one thing that I think would be relevant to ask would be: What impact is it going to have on the general economics of the dairy industry? What impact is it going to have on communities that rely upon dairy economics? There may be risk issues, safety issues, and health issues in terms of the family farm.

Martin: OK, so economic issues as well as the science behind bST are important. Now let me change my assumptions just slightly. I am going to come back to both of you with a question. The new information is that FDA approved milk from cows treated with bST as safe about six years ago, but has been analyzing its potential impacts on dairy cows and the environment. The new news story is that bST has just been approved by FDA for commercialization. You must write a story on this news event. Karen, and then David, how would you approach the story of FDA approval of bST for commercial use?

Bolluyt: That is difficult from a university perspective, since it is basically a private-sector story. What we might do is contact our scientists who have expertise in the area and put together a contact list for reporters.

Martin: So you would provide a source of information for reporters who are writing for various national or local newspapers as to where they might go for background information or answers to questions.

Bolluyt: Right.

Martin: David, how would you approach this bST story?

Judson: I don't know. There is news value in the symbolic crossing of the regulatory threshold by one company. In other words, is this a harbinger of things to come? I think this would be the dimension of the story that would occur to me most immediately.

Martin: Ann Sorensen, let me turn to you for a moment. I want to bring the farmers into the discussion. FDA has now approved bST for at least one of the companies. You work with farmers. How do you think farmers might respond to this news?

Sorensen: I wish I had the answer to that. I think farmers are split on the issue of bST. I know some dairy farmers who are very concerned about negative consumer reactions. There will probably be some dairy farmers who can incorporate the product immediately into their operations if they feel it is going to increase their efficiency or their productivity. I think there are other dairy farmers who won't touch it. They either will not see a need for it or will be concerned about negative consumer reactions.

Martin: So there may be differences depending on a farmer's operation, economic situation, ability to manage the technology, and how he markets his milk.

Sorensen: Right. They are going to have to manage the new technology. They are going to have to keep track of individual cattle. They are going to have to know what their cows are eating. It is going to require, I think, a high level of management. Not all dairy farmers can do that.

Martin: Rosetta, you work with the food industry and food processors. How would this news event impact some of the people you work with?

Newsome: First, let me clarify that while many of the members of the Institute for Food Technologists are employed in the food industry, others are employed in regulatory agencies and academia. Perhaps, my primary role might be to provide the scientific information for inquiries we get, whether they be from journalists or consumers. If safety concerns remain, then we would strive to provide the background information on the science to satisfy these concerns.

Martin: Rebecca Goldburg, I would like to come back to you for a minute. I know you work closely with some of the environmental organizations. Would this news about FDA approval of bST raise questions or concerns from an environmental perspective?

Goldburg: I actually have not worked on any issues related to bGH (bovine growth hormone); so my concerns would probably be of a personal nature, and perhaps less about the safety of the milk and more about how we treat animals.

Martin: So you would be more concerned about animal welfare, animal management, or animal well–being.

Goldburg: Right, and also economic issues.

Martin: OK, but what about beneficial or adverse environmental impacts?

Goldburg: Well, there could be some adverse environmental impacts if we lose a lot of dairy farms in the northeast. This raises land management issues. I see it not so much as an environmental issue, but more of an agricultural policy and consumer safety issue.

Martin: Let me turn back to the audience. What we have done tonight, I hope, is give you a flavor of NABC 5. I hope we have set the tone for the conference with two or three examples of some of the kinds of issues of concern to society, to the organizations that we represent including the private sector, to academia, to public interest groups, and to regulatory agencies. Some of the biotechnology examples we discussed this evening are already in place, some are on the near horizon, and others may be further away. Each of you has selected one of four workshops. The four workshops are: public values, public assessment of biotechnology, technical assessment of risks and benefits associated with biotechnology, and, finally, communication about risks. You will have six hours of workshop sessions, tomorrow and on Friday, where you can speak, share your views, raise your questions, express your concerns, or make your recommendations. I hope you do that. I hope you participate. In closing, let us take a moment to express our appreciation to our panel members. Thank you very much. I hope this was an enjoyable evening for all of you. Good night. (Applause).

Participants

Roger Balk
Royal Victoria Hospital
687 Pine Avenue West
Montreal, Que., Can. H3A 1A1

Patrick Basu
USDA
14th and Independence
Washington, DC 20250

Bill R. Baumgardt
Purdue University
116 AGAD
West Lafayette, IN 47907

Robert Benson
4265 E. 169th St.
Noblesville, IN 46060

Jeff Bergau
1033 University Place, #450
Evanston, IL 60201

Calvin Bey
Box 96090
Washington, DC 20090

Christopher Bidwell
An. Sci., Purdue University
1026 Poultry Sci. Bldg.
West Lafayette, IN 47907

Jerry Bishop
Wall Street Journal
200 Liberty Street
New York, NY 10282

Karen Bolluyt
Iowa State University
304 Curtis Hall
Ames, IA 50011

Karl Brandt
Dept. of Agriculture
Purdue University
West Lafayette, IN 47907

Christine Bruhn
Center for Consumer Research
University of California
Davis, CA 95616

J. Bruce Bullock
MO Ag Exp Stn, Coll. of Ag.
2–64 Agriculture Building
Columbia, MO 65211

Bees Butler
Dept. of Ag Economics
University of California
Davis, CA 97313

Kay M. Carpenter
1645 W. Valencia Drive
Fullerton, CA 92633

Carolyn Carr
University of Minnesota
100 Ecology Bldg.
1987 Upper B Circle
St. Paul, MN 55108

William E. Chaney
U of CA, 118 Wilgart Way
Salinas, CA 93901

Rebecca Chasan
15501 Monona Dr.
Rockville, MD 20855

Alfred Chiscon Jr.
Purdue University
West Lafayette, IN 47907

Ronnie Coffman
Cornell University
248 Roberts Hall
Ithaca, NY 14853

Réne Colon
Dept. of Agriculture
504 Barbe Street
Santurce, Puerto Rico 00912-3912

Robert Gast
Michigan State University
Ag. Exp. Station
109 Agriculture Hall
East Lansing, MI 48824-1039

Ian Gray
Michigan State University
Ag. Exp. Station
109 Agriculture Hall
East Lansing, MI 48824-1039

Ralph Hardy
Boyce Thompson Institute
Tower Road
Ithaca, NY 14853-1801

Susan K. Harlander
Land O' Lakes, Inc.
PO Box 116
Minneapolis, MN 55440

Bud Harmon
Dept. of Animal Sciences
Purdue University
West Lafayette, IN 47907

Betsy Hill
The NutraSweet Company
1751 Lake Cook Road
Deerfield, IL 60015

John Hitchell
1014 Vine Street
Cincinnati, OH 45202

Laura Hoelscher
Ag. Communication Service
Purdue University
West Lafayette, IN 47907

Wayne Dillman
3901 W. 86th St., #285
Indianapolis, IN 46268

Peter Dunn
Agricultural Biotechnology
Purdue University
1158 Entomology
West Lafayette, IN 47907-1158

Sharon Dunwoody
School of Journalism
University of Wisconsin
821 University Ave.
Rm. 5115
Madison, WI 53706

Stan Ernst
2021 Coffey Road
Columbus, OH 43200

Dennis Erpelding
Elanco Animal Health
1901 L St. NW, #705
Washington, DC 20036

Will Erwin
2595 14B Road
Bourbon, IN 46504

Walt Fehr
Iowa State University
1010 Agronomy
Ames, IA 50011

Richard Feinberg
Consumer Sciences
Purdue University
West Lafayette, IN 47907

Mike Ferring
3800 Barham Blvd. Ste. 409
Los Angeles, CA 90068

Joseph Fordham
Novo Nordisk Bioindustrials, Inc.
33 Turner Road
Danbury, CT 06813–1907

Roy Fuchs
Monsanto Company, GG4G
700 Chesterfield Village Pkwy.
St. Louis, MO 63198

Jennifer Garrett
University of Missouri
5102 An. Science Center
Columbia, MO 65211

Rebecca Goldburg
Environmental Defense Fund
275 Park Avenue South
New York, NY 10010

Alan Gould, Director
Biotech. and Plant Genetics
DowElanco
9410 Zionsville Road
Indianapolis, IN 46268

Wm. Greenlee
Dept. of Pharm. and Toxicology
Purdue University
West Lafayette, IN 47907

Eddie Hansen
MSU, Ag. Exp. Station
109 Agriculture Hall
East Lansing, MI 48824-1039

Sarah Hanson
Int'l. Food Information Council
1100 Connecticut Ave., NW #430
Washington, DC 20036

Richard Harwood
Michigan State University
A260 Plant and Soil Science
East Lansing, MI 48824

Mark Henderson
DowElanco
9002 Purdue Road
Indianapolis, IN 46268

Mike Hendricks
Kansas City Star
1729 Grand Avenue
Kansas City, MO 64108

Thomas J. Hoban
Depts. of Sociology & Anthropology
NCSU, Box 8107
Raleigh, NC 27695-8107

Theodore L. Hullar
U of CA., Davis
567 Mrak Hall
Davis, CA 95616

Elke G. Jarchow
Ciba-Geigy Ltd.
CH 4002
Basel, Switzerland

Karla Spencer-Johnson
University of Missouri
5102 An. Science Center
Columbia, MO 65211

Dawn Jones
Stephan & Brady
1850 Hoffman Street
Madison, WI 53704-2594

David Judson
Gannett News Service
1000 Wilson Blvd.
Arlington, VA 22229

Dave King
Agricultural Communications
1143-AGAD
Purdue University
West Lafayette, IN 49706

Janet Klein
WDATCP
801 West Badger Road
Madison, WI 53713

Brewster Kneen
125 Highfield Road
Toronto, Ont., Can. M4L 2V4

Desmond Jolly
Agriculture Economics
U of CA, Davis
Davis, CA 95616

Brent Ladd
Dept. of Animal Sciences
Purdue University
West Lafayette, IN 47907

Victor L. Lechtenberg
Ag. Exp. Stn., 114 Ag. Admin.
Purdue University
West Lafayette, IN 47907

George Lee
University of Saskatchewan
204 Kirk Hall
Saskatchewan, Sask., Can. S7N 0W0

June Fessenden MacDonald
NABC, 159 Biotechnology Bldg.
Cornell University
Ithaca, NY 14853-2703

David R. MacKenzie
Cooperative State Res. Serv.
USDA CSRS, Rm. 330
Aerospace Center
Washington, DC 20250-2220

Roger Maickel
Dept. of Pharmacology
Purdue University
West Lafayette, IN 47907

Marshall Martin
Ctr. for Ag. Policy and Tech. Transfer
Purdue University
1145 Krannert Bldg.
West Lafayette, IN 47907

Barbara Masters
PO Box 1327
Harpers Ferry, WV 25425-1327

Ted McKinney
DowElanco
9002 Purdue Road
Indianapolis, IN 46268

Murray McLaughlin
Ag-West Biotech Inc.
304-111 Research Drive
Saskatoon, Sask., Can. S7N 3R2

Laura Meagher
Ag. Biotech. Center
Rutgers University
PO Box 231, Cook College
New Brunswick, NJ 08903

Libby Mikesell
Int. Food Information Council
1100 Conn Ave., NW # 430
Washington, DC 20036

Joseph Miller
Indiana Farm Bureau
PO Box 1290
Indianapolis, IN 46206

Scott Mills
Dept. of Animal Science
Purdue University
West Lafayette, IN 47907

Kamendra N. Mishra
Bioseed Genetics Intl.
2010 S. Ankeny Blvd.
Ankeny, IA 50021

R.E. Morgan
Saskatchewan Wheat Pool
Box 1461
207-112 Research Dr.
Saskatoon, Sask., Can. S7K 3P7

Anne Mueller
PO Box 751
Wilmington, DE 19897

John Neilson
1022 McCarty Hall
University of Florida
Gainesville, FL 32611

Rosetta Newsome
Int'l. Food Technologists
221 N. LaSalle St
Chicago, IL 60601

Kate O'Hara
NABC,159 Biotechnology Bldg.
Cornell University
Ithaca, NY 14853-2703

Eldon E. Ortman
Ag. Exp. Stn. / 122 AGAD
Purdue University
West Lafayette, IN 47907

Raymond Ortman
PO Box 745
Kokomo,IN 47971

Hamer Paschal
Asgrow Seed Co.
205 N. Michigan
Oxford, IN 47971

Harry L. Pearson
Indiana Farm Bureau
130 East Washington Street
PO Box 1290
Indianapolis, IN 46206

Sandra R. Rathod
Consumer Family Sciences
Purdue University
West Lafayette, IN 47907

Keith Redenbaugh
Calgene, Inc.
1920 5th Street
Davis, CA 95616

William Reinert
Campbell Inst. for Res. & Tech.
Route 1, Box 1314
Davis, CA 95616

Rex Ricketts
5102 Animal Sci. Center
University of Missouri
Columbia, MO 65211

Max Rothschild
Ag. & Home Ec. Exp. Stn.
18 Curtiss Hall
Ames, IA 50011

Nilda Pérez-Roussett
PR Dept. of Agriculture
8 St D 25, Parque De Joromar
Bayamon, Puerto Rico 00959

Lilly–Marlene Russow
Dept. of Philosophy
Purdue University
West Lafayette, IN 47907

Eric Ryden
2343 Greenwood Ave.
Wilmette, IL 60091

Allan Schmid
Michigan State University
36 Agriculture Hall
East Lansing, MI 48824

Karen Schneider
U of VT. Extension
RR 4, Box 1308
Middlebury, VT 05753

Norm Scott
Cornell University
314 Day Hall
Ithaca, NY 14853

John R. Snyder
Office of Tech. Transfer
Purdue University
West Lafayette, IN 47907

A. Ann Sorensen
American Farmland Trust
Ctr. for Ag. & the Environment
PO Box 987
Dekalb, IL 60115

John Sorenson
7000 Portage Rd.
Kalamazoo, MI 49001

Patrick Stewart
American Farmland Trust
PO Box 987
DeKalb, IL 60115

Kay Shipman-Swiech
1701 Towanda Ave.
Bloomington, IL 61701

J. Scott Thenell
DNA Plant Technology
6701 San Pablo Ave.
Oakland, CA 94608-1239

Michael Thomashow
Michigan State University
A291 Plant and Soil Sciences
East Lansing, MI 48824

Paul Thompson
Ctr. for Biotech. Policy and Ethics
Texas A & M Univ., 329 Bell Building
College Station, TX 44843-4461

Frank Thorp
Thorp Seed Company
Route 3, Box 257
Clinton, IL 61727

Rodney Tietz
Asgrow Seed Company
205 N. Michigan
Oxford, IN 46268

Patricia Traynor
Michigan State University
110 Biology Res. Center
East Lansing, MI 48824-1320

Richard Traxler
Dept. Food Science
University of Rhode Island
530 Liberty Lane
W. Kingston, RI 02892

Diane van Loan
3165 McKelvey Rd., #240
St. Louis, MO 63044

Connie Vidos
MBD
1100 Connecticut Ave., NW
Suite 300
Washington, DC 20036

Jane Voichick
Nutritional Sciences
U of Wisconsin, Madison
1415 Linden Drive
Madison, WI 53706-1571

Robert D. Waltz
IN Dept. of Nat. Resources
613 State Office Building
Indianapolis,IN 46204

John Warner
3800 Barham Blvd., #409
Los Angeles, CA 90068

Don Weeks
Center for Biotechnology
University of Nebraska
101 Manter Hall
Lincoln, NE 68588–0159

Scott Whitman
Ag. Engineering
Purdue University
West Lafayette, IN 47907

Rob Ziegler
600 W. Main St.
Ft. Wayne, IN 46802

NOVARTIS
CROP PROTECTION
Research Library

WITHDRAWN
FROM LIBRARY

MAR 1 5 1999

NOVARTIS
CROP PROTECTION